MYTH BUSTING
PHYSICS
THIRD EDITION

ROGER I. PARKER II

Archway Publishing books may be ordered through booksellers or by contacting:

Archway Publishing
1663 Liberty Drive
Bloomington, IN 47403
www.archwaypublishing.com
844.669.3957

Interior Image Credit: Roger I Parker II

ISBN: 978-1-4808-9747-2 (sc)
ISBN: 978-1-4808-9748-9 (e)

Library of Congress Control Number: 2020920118

Print information available on the last page.

Archway Publishing rev. date: 11/06/2020

Contents

INTRODUCTION TO 2020
THIRD EDITION

I started this project in 2009, when I noticed a mystery of science. Who doesn't love to solve a mystery, right? I came across an article in **Science** magazine from January 4, 2008 which described four–dimensional spatial chemical reactions in crystals. [1]

This goes against what Albert Einstein said in his General Theory of Relativity. In his theory, he stated that there are only three spatial dimensions and a single time dimension, and when they are combined make up what he calls a space-time continuum. Nevertheless, this article claims that there are four spatial dimensions and it does not refer to a time dimension. Who is right: Einstein or the 2008 **Science** magazine article?

I also came to question Einstein's claim that "time" was some sort of dimension. After all, what does any clock measure? The most advanced clock ever made – the atomic clock – measures the vibrations of a quartz crystal and synchronizes it with the vibrations of a cesium atom. A definition of one second is 9,192,631,770 oscillations (vibrations – or movements) of a cesium 133 atom. According to John Langone, the international definition for a second according to The International Committee on Weights and Measures is defined as "the time an electron takes to spin on its axis inside a cesium 133 atom; and this translates to 9,192,631,770 oscillations." [2]

Also, is there any experimental evidence that "time" is any sort of dimension? I have never seen any so far.

Therefore, the definition of "time" must be tied to the synchronized movements of two objects in space. "Time" is movement. And our sense of "time" is tied to our sense of motion. All the definitions for our measurements of "time" are tied to some type of motion. A year is defined as one full orbit of our planet around the sun. A month and a week are a partial orbit of our planet. A day is defined as Earth making one full spin on its North – South Axis. Both a minute and an hour can be measured and defined in many different ways. One way of defining a minute and hour is to say that they are some percent of the Earth's motion as it spins on its axis. "Time" can only be defined as a motion in space. Whether that space in which this motion occurs is one, two, three or four –dimensional is another issue altogether – see Figure 1. I explain more in Section 2 of this book.

The Definition of	is	motion	object
a year	is one	orbit	of Earth around the sun
a month	is a partial	orbit	of Earth around the sun
a day	is one	rotation	of Earth on its axis *
an hour	is a partial	rotation	of Earth on its axis *
a minute	is a partial	rotation	of Earth on its axis *
a second	is 9,192,631,770	oscillations	of a cesium 133 atom
Time	is a	motion	through a given space

* A 'day' can also be mesured by the path of a swinging pendulum.

FIGURE 1

This posed a mystery. If the information in the article was true, then why was this not front–page news: "Einstein Was Proven Wrong!" We apparently live in a universe that has four physical spatial dimensions and not three as Einstein claimed. Why hadn't I heard about it before? And why did the article present the material as common knowledge?

I wanted to tackle this mystery. Luckily, I had collected books since I was 12, so by the time I started investigating this mystery of science I had a fully stocked library to work with.

In 1980, there was a PBS television series called **Cosmos** that I loved. [3]

My mom and dad had bought me the **Cosmos** companion book to the television show for Christmas, and it is one of my most prized possessions. My favorite episode is "The Edge of Forever." In this episode, Carl Sagan said that our universe maybe a four–dimensional sphere – a hypersphere. This always puzzled me because if this is the case then why is our universe just composed of three–dimensional bits – (three–dimensional galaxies)? Chemistry books do not depict atoms as being two–dimensional. If I pick up a ball, I know that the ball itself is composed of three–dimensional atoms. So why should the universe be any different? Why is the four–dimensional universe composed of three–dimensional atoms instead of four – dimensional atoms?

Maybe, the 2008 **Science** magazine article is correct. Maybe, there is a four–dimensional atomic structure to our universe. If that is true, two questions emerge. First, can an experiment prove the existence of a four–dimensional atomic structure? I discovered the answer to this question is, yes (see Section 1, Part 5). Second, why can't we fly off into the fourth dimension? (The answer to this question is in Section 2, Part 2, Proof1).

A couple more questions need to be asked: 'why did Einstein insist that there are only three spatial dimensions to our universe when crystals are shown to have four spatial dimensions in their atomic make up?' Why has this issue been ignored for so long?

In 1905, when Einstein wrote his paper on the General Theory of Relativity, his equations called for an ever–changing universe that was either expanding or contracting. Therefore, Einstein asked astronomers whether the universe was expanding or contracting. At the time, however, astronomers considered the universe to be our galaxy and nothing more. It was thought that other galaxies might exist, but this had never been proven. So, in 1905, the astronomers told Einstein that the universe was static, it never changed.

After that, Einstein added a new number to his equation and called it the Cosmological Constant. Then in 1929, Edwin Hubble published a paper that proved the existence of other galaxies and it showed that the universe was expanding. Einstein would later say that putting the Cosmological Constant into his original equation was had been his "biggest blunder." However, in 1905, nobody knew that there were galaxies other than our own, nor could they have known what those galaxies were doing. New scientific discoveries are made every day. So, I wondered, was Einstein also wrong about there being only three spatial dimensions?

I remembered another of my library's treasures, a book called **The Fourth Dimension: A Guided Tour of the Higher Universes,** by Rudy Rucker. In his book, Rucker describes how physicists and mathematicians during the years between the 1800's and the 1920's believed that there might be four spatial dimensions in our universe. [4.1, 4.2]

One of these scientists was named William Kingdon Clifford. Some scientists feel that he is the true father of relativity, because Einstein used many of his ideas. [5.1 – 5.3]

One of Clifford's ideas was the geometric theory of gravity. He wrote an essay, **On the Space – Theory of Matter,** in 1870. In this work, Clifford explained that energy and matter are different types of spatial curvatures. These two ideas were crucial in helping Einstein develop his theory of relativity. William Clifford was trying to develop relativity in 1870, but he died suddenly in 1879, leaving his work on relativity incomplete. Ironically, Einstein was born the same year that Clifford died. Einstein would pick up where Clifford left off, but Einstein did not take relativity in the same direction Clifford would have. [6]

One of the fundamental differences between Clifford's relativity and Einstein's relativity was that Clifford's relativity had three spatial dimensions curved into a fourth spatial dimension. Einstein was probably skeptical of the idea of a fourth spatial dimension. After all, what would possibly prevent someone flying off into the fourth spatial direction? That is most likely why Einstein chose **TIME** as the fourth dimension. After all, you cannot fly off into time.

In 1905, there was no proof of a fourth spatial dimension to our universe. However, during the 1920's, that would change. New evidence emerged to show that our universe might indeed have four spatial dimensions, and that evidence came when geologists took x–rays of crystals to determine their structure. When the geologists did this, they noticed structures that could only occur if there were four spatial dimensions to our universe.

With this discovery, it should have been obvious that Einstein made a second "blunder" by stating that there were only three spatial dimensions with time as a non–spatial dimension. To my knowledge, nobody, besides me, has ever caught this glaring error. Skepticism can be a double—edged sword in that Einstein did everything right. He took the right approach in being skeptical of his own equations, which prompted him to ask astronomers of his day what the universe was doing, and their response prompted Einstein to then modify his original equation accordingly. When new evidence became known, proving the astronomers and Einstein wrong, it shows us that even if you do everything right, new evidence can prove you wrong. This is how our scientific knowledge increases.

In the First Part of the First Section of this book, we will look at skepticism, which you have already seen, can be a double—edged sword. We will skeptically examine why some physicists believe in the myth that they can get something from nothing. In the Second Part of the First Section, I show how the Pauli Exclusion Principle provides a way to either prevent time travel or to make time travel very difficult. In the Third Part of the First Section, we will see how photons have mass. In the Fourth Part of the First Section, I cover how the Casimir Effect is really caused by thermal fluctuations in a vacuum. Then we will look at how this is connected to the large-scale structure of our universe, and how it operates. Next, we discover that temperature fluctuations cause the warping of space. By revealing this cause – effect relationship between thermal radiation (light) and the warping of space (gravity) we have achieved the Grand Unified Field Theory that Einstein was after. Then I show how this Grand Unified Field Theory can be experimentally proven in the Fifth Part of the First Section. In the Sixth Part of the First Section I tie up many loose ends in this section of the book.

In the First Part of the Second Section of this book we first look at how temperature fluctuations and not mass determine the fate of the universe. In this part of the book we start to look at the possibility that our universe is indeed four-dimensional with four dimensional structures like Cosmic Void Spikes. Then we look at how our universe has four spatial dimensions and how energy and matter are distributed in these four spatial dimensions. Also, the definition of what is meant by "universe" is presented in this section and how this relates to what cosmologists call the Cosmic Box. Then we briefly look at the evolution of the concept of "universe". We also look at the possibility that the speed of light is variable. In the Second Part of the Second Section in the First Proof we will look at how the accelerated expansion of our universe prevents us from moving in the fourth spatial dimension. To understand this, we must look at what Newton said about acceleration. Newton stated that anything in our universe that accelerates produces g—force and it is the g—force our universe produces as it accelerates that prevents us from ever moving in the fourth spatial dimension. Then we discuss the Cosmic Horizon Problem. In the Second Part of the Second Section in the Second Proof the four—dimensional objects called Hyper Holes are described. Hyper Holes are not just another name for "black holes". Black holes are singularities. Hyper Holes are not singularities. Hyper Holes are merely four—dimensional holes. Is the hole in your kitchen sink a singularity? No. Hyper Holes are like the hole in your kitchen sink in that they are not singularities either. Also, the mass and weight of all holes is by definition zero. After all, talking about the "mass" and / or "weight" of a 'black hole' makes as much sense as asking about the "mass" and / or "weight" of the hole in your kitchen sink! It makes no sense at all because in both cases how would you even be able to determine the "mass" of either hole? For example, how would you measure the mass and / or weight of the hole in your kitchen sink? Would you measure the mass and weight of the kitchen sink, after all, the hole does go through the kitchen sink? The problem with this is that you are measuring what the hole goes through but you are not measuring the hole itself. What if you measure the mass and / or weight of the air that fills the hole in an empty kitchen sink? You are still not measuring the mass and / or weight of the hole; you are only measuring what is filling the hole. A singularity is like a blockage in the pluming of a kitchen sink so if you measure the mass and or weight of it you still will not be measuring the hole itself; you will instead only be measuring the properties of what is blocking up the hole. Einstein got it right when he said that singularities are impossible to make. I explain all of this in more detail in Section 2, Part 2, Proof 2. [7]

It is these Hyper Holes that drive the expansion of our universe. It is the formation of a Hyper Hole that gave birth to this universe and other universes are formed in the same way. I then talk about how Hyper Holes are like Klein Bottles.

In the Second Part of the Second Section in the Third Proof brings us to a discussion of the four-dimensional crystals to show that Dark Matter and Dark Energy are the ordinary four-dimensional matter and energy that makes up our universe. We cannot see or move in the fourth dimension because the g-forces caused by the expansion of universe prevents it. The four-dimensional structure of crystals proves that our universe is four-dimensional down to its atomic structure, and I describe a way to prove that the crystals are really four-dimensional. Then I describe what "time" is. Next, I talk about "faster-than-light" communication. Then in Section 2, Part 2, Proof 5 I discuss how gravitational lenses prove that space is curved into a fourth spatial dimension. In Section 2, Part 3 and Part 4 the next topic I discuss is the two other ways in which time travel would be very difficult to achieve. Next, I give a better explanation of Enthalpy. The last topic in this section I discuss is how thermodynamics (infrared light) does work on our universe by warping space as the universe expands.

In the Third Section of my book I will talk about the applied science – the technology that will come from my work. The first technology that I will discuss is warp drive technology. The second technology I will talk about is four – dimensional communications.

Using classical physics and four-dimensional geometry I disprove the following myths:

1. Time is the fourth dimension
2. Quantum fluctuations in a vacuum exist
3. Photons have no mass
4. There is nothing outside of our Observable Universe
5. Singularities exist
6. Nothing is colder than Absolute Zero

YOU CANNOT GET SOMETHING FROM NOTHING

PART 1

PROBLEMS WITH MODERN PHYSICS

This section of the book is titled "You Cannot Get Something From Nothing," which you would think is self-evident in our everyday reality, right? So, why do some modern physicist, like Brian Greene, insist that you **CAN** get something from nothing?

In fact, this is one of the most pernicious myths in physics. Many theories of physics count on you being able to get something for nothing. Many physicists place this assumption as a cornerstone of their theories. I feel that nothing should be taken for granted or assumed out of hand, and that the role of science is to confront and challenge such assumptions. We will explore this assumption in this book and other such assumptions.

Physicists, such as Roger Penrose recognize that fashion, faith, fantasy, and flights of fancy all play a part in many of the theories of modern physics. Roger Penrose believes that if a theory can be shown to be influenced by the enslavement of fashion, by an unsupported faith, or by some fantasy, then it is the role of science to point out these erroneous influences and steer the public from influences of this kind. [8]

Astrophysicist Neil deGrasse Tyson is highly skeptical of some of the modern theories of physics, such as string theory. [9]

Many others are also skeptical of modern theories of physics. For example, Tim Folger states that a dozen interpretations of quantum mechanics compete for the attention of physicists. Each of these interpretations of quantum mechanics has a radically different take on reality. Adan Cabello, a physicist at the University of Seville in Spain, has summed up the confusing and incompatible gaggle in viewpoints of quantum mechanics as a "map of madness". [10]

Folger says: "If physicists can't agree on – or don't know – what their reigning theory is all about, does it mean they've hit a wall in terms of understanding the world?" [11]

Roger Penrose named his book ***The Emperor's New Mind*** after a children's fairy tale called the Emperor's New Clothes. Was Roger Penrose implying that physicists of today are in some ways like the Emperor in the children's fairy tale? In the fairy tale the Emperor did not have new clothes he had no clothes, so is Roger Penrose saying that the physicists of today do not have a new mind but instead no mind? I guess what he means to say is that there are many mysteries of the universe that leave physicists totally clueless. [12]

Neil deGrasse Tysons feels that people should learn to think for themselves by questioning authority when necessary. He believes that, no idea is true just because someone in authority says so. However, he also feels that you should question yourself because you should not believe something just because you want to. Believing in something does not make it so. Tyson wants people to be more scientifically literate in general, at least enough to test ideas using evidence gained by observation and / or experimentation. When a scientist performs a well-designed test and it proves all his ideas wrong, they should just get over it. Science is all about following the evidence, wherever it leads. If a scientist has no evidence, then they should reserve judgement. They should always remember that they could be wrong. [13.1 – 13.2]

Question dogma.

In an interview with Larry King, Neil deGrasse Tyson stated that everyone should be a self – driven learner. Neil deGrasse Tyson thinks that people in general need to be more curious about things they don't know. He wants them to investigate things that make no sense to them. [14]

That is partially what this book is about. This book is my attempt to make physics more understandable and to investigate the things in physics that puzzle us all – layman and scientist alike.

Massimo Poldoro says that "the words of an authority have no value unless they are supported by facts." [15]

Right now, modern physics is divided into three distinct fields: quantum physics, classical physics, and cosmological physics. Many physicists have been looking for a grand theory to unify these three separate branches of physics. **Quantum physics** describes the world of the minuscule (atoms and subatomic particles), **classical physics** describes the world of our everyday reality (chemical reactions of gases, liquids and solids it also describes how mechanical devices work), and **cosmological physics** describes the world of the enormous (the motion of the stars, planets, and galaxies).

Each field of physics has little interaction with the others and operates under different laws, rules, and assumptions. For example, in classical physics there is a set of conservation laws stating that you cannot get something from nothing.

However, both quantum physics and cosmological physics agree that you **CAN** get something from nothing. For example, cosmological physics states that our universe came from nothing and that it suddenly exploded into being. Quantum physics has as its bedrock subatomic particles popping into existence then winking out in a vacuum, which is a clear violation of the laws of conservation in classical physics both coming and going. Both quantum physics and cosmological physics are in direct contradiction with this and other issues in classical physics. Thus far, only classical physics has given us experimental and observational evidence for its claim that you cannot get something from nothing. Neither quantum physics nor cosmological physics have produced any credible experimental or observational evidence that you **CAN** get something from nothing. There is no experimental proof that the First Law of Thermodynamics is invalid.

In this Section of the book, I will attempt to apply the well–proven law of classical physics – that you cannot get something from nothing – to the extremes of scale, in quantum physics (small) and cosmological physics (large). I will give logical alternative perspectives to many theories in modern physics today and prove that you really do not get anything from nothing. In this way, I will unify all of physics under one rule regardless of the scale, so each field of physics follows the same rules. Wishing to receive something from nothing is a classic case of having one's cake and eating it too.

PART 2

PAULI EXCLUSION PRINCIPLE

I have a piece of cake on a plate in front of me, and the rest of the cake is over there on another plate. It is an indisputable fact that I can either eat this piece of cake and have it digesting nicely in my tummy, or I can have it on my plate in front of me, but both events cannot happen at the same time because of the Pauli Exclusion Principle.

The Pauli Exclusion Principle states that fermions (electrons, protons, neutrons) cannot be in two different places with the same energy at the same time. [16.1, 16.2]

This principle governs what happens in a white dwarf star, namely ensuring that no two electrons can occupy the same space. This is what creates the electron degeneracy pressure that holds the star up against further gravitational collapse. [17]

Therefore, the atomic structure of an individual slice of cake can be either on my plate or in my tummy, but that particular slice of cake cannot be in two places at once, according to the Pauli Exclusion Principle. I can eat one slice of cake here and then put a **different slice of cake** on my plate, but I cannot have the same slice of cake in two places at once – on my plate and in my tummy.

The Pauli Exclusion Principle suggests that each atom that makes up an object (a piece of cake, a bar of iron, a human) is unique and no such object can be made of a single atom whizzing around, nor can an object be made up of identical atoms in identical states. All objects in our universe are composed of collections of unique atoms. This means that the Pauli Exclusion Principle will prevent Schrödinger's Cat from ever being in two states or positions at the same time before and or after you look in the box.

It is not hard to understand how atoms, like snowflakes, can be unique, or one of a kind. [18]

Each individual snowflake forms under unique conditions of temperature and humidity, which makes each snowflake a unique crystal structure. Could the atoms form from quarks in the same manner, as snowflakes making each atom, though similar a unique individual?

Carl Sagan and Benoit Mandelbrot have suggested that our universe is probably a fractal universe. [19.1 – 19.4]

Similar ideas were expressed by Thomas Wright and Immanuel Kant. [20]

In, **Cosmos** Episode 10: "Edge of Forever," Carl Sagan described one of the strangest and most haunting ideas in science: A Fractal Cosmos as an infinite hierarchy of universes. What this means is that every elementary particle in our universe, such as an electron, would reveal itself to be another closed universe if we could investigate it. If that were true, this would mean that every electron in our universe is its own universe – with galaxies, stars, and electrons that are in turn even smaller universes and so on. There is an infinite regression of electron–universes downward. Sagan describes an infinite regression upward as well, where our universe may be just an electron in a much larger universe, and that universe is just an electron in an even larger universe, and so on upward into infinity. An infinite regress of universes up and down. This is an idea that in Sagan's words, "stirs the blood." [21]

If Carl Sagan and Benoit Mandelbrot are correct in saying that we live in a fractal universe, then it is conceivable that even though each universe – (like each snowflake in a group of snowflakes) – is similar, each universe is also unique, having formed under unique conditions. [22.1 – 22.4]

Another area affected by the Pauli Exclusion Principle is time travel. Some aspects of time travel theory depend upon the magical conjuring of matter from nothing. These aspects of the theory of time travel not only violate the Pauli Exclusion Principle but they also violate the First Law of Thermodynamics, which states that you cannot make or destroy energy. These same aspects of time travel theory also violate Einstein's equation $E=mc^2$, which tells us that energy and matter are proportionally related to one another. Therefore, if you cannot make nor destroy energy, you cannot make nor destroy matter as well.

One such example of how time travel is affected by the Pauli Exclusion Principle is given by J. Richard Gott. Gott had a former student calculate a vacuum state in a basic spacetime involving time travel, which he calls the Groundhog Day Spacetime. Gott uses the movie **Groundhog Day** as a model for this spacetime. In the movie Bill Murray's character relives the same day over and over, which happens to be Groundhog Day. When Bill Murray's character goes to bed each night, he sleeps until his alarm sounds at 6:00 A. M. on Groundhog Day. He is dismayed when he discovers that he is about to relive the events of Groundhog Day again. Richard Gott's Groundhog Day spacetime is formed by taping 6:00 A. M. Tuesday to 6:00 A. M. Wednesday. In this Groundhog Day spacetime line when you reach the end of the day on Wednesday, you automatically wake up at 6:00 A. M. on Tuesday to begin the cycle all over again. Richard Gott envisions this cyclical timeline as a helix wrapped around a cylinder. He believes that if you lived for 80 years (29,220 days) in this Groundhog Day spacetime, your timeline would encounter 29,219 copies of you, ranging in age from babies to senior citizens. [23]

Okay, so where do all these copies of yourself come from, and especially where do the babies come from? Who gives birth to the baby copies of you? The Pauli Exclusion Principle says the matter that makes up your body can only be in one place at any given time. Einstein's equation $E=mc^2$ says that neither matter nor energy can come from nothing, nor can matter or energy be destroyed. This equation supports and confirms the First Law of Thermodynamics, which states that energy cannot be made nor destroyed. More importantly, where does all the matter and energy needed to make the copies of you come from? The atoms within you can only be inside you **OR** the copy of you; they cannot be in two places at once. All of this adds up to the fact that the other copies of yourself can never form or exist. There can only be one you period.

Stephen Hawking proposed a Chronology Protection Conjecture which says that the laws of physics conspire to prevent time travel from happening. For more information on other problems with time travel look at Parts 3 and 4 in the Second Section of this book. [24]

The Pauli Exclusion Principle, the First Law of Thermodynamics, and Einstein's equation $E=mc^2$ certainly prevent J. Richard Gott's Groundhog Day from ever happening. However, what if I wanted to visit my great, great, great grandmother or a mammal ancestor that lived at the time of the dinosaurs? In these cases, all three rules of physics also apply to prevent me from going back through time, because there is an unbroken chain of life. You see there is a certain amount of matter that was once part of my mother and / or animal ancestor but is now in me – about 10 pounds is transferred from mother to child across time. Over the gestation period about 10 pounds is transferred from my mother to me. I was once an egg inside my mother, then I grew to about 10 pounds in her womb – in other words, about 10 pounds of my mother became me. Therefore, there is an unbroken string of matter that links all life to the past – about 10 pounds' worth. It is these 10 pounds that prevents me from going back through time because the 10 pounds that was in my mother and that was in my great, great, great grandmother can be either in me or in my great, great, great, grandmother but it cannot be in two places at once.

Gott's assumption that copies of you would miraculously form from thin air in the Groundhog Day space-time line is faulty at best. All the atoms within you can only be in you **OR** in a copy of you, but they cannot be in both places at the same time. This means that matter and energy would have to come from somewhere else to make even a single copy of you in the Groundhog Day spacetime line. It would amount to a free lunch – getting something for nothing. And that would be a direct violation of the conservation of energy and matter, as in $E=mc^2$. Remember, you cannot have your cake and eat it too.

Let us do a thought experiment to better illustrate a similar problem with time travel. In this thought experiment, let's say we have a cat named Angelica, and you want her to meet her great, great, great, great, great grandmother Stephany.

In this thought experiment all the cats live to be 50 years old and they have kittens when they are 30 years old. See Figures 2 – 4.

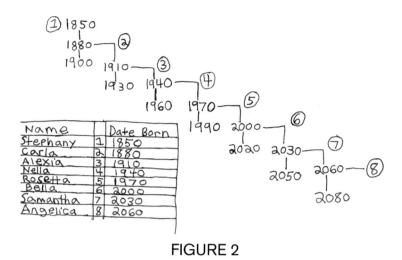

Name		Date Born
Stephany	1	1850
Carla	2	1880
Alexia	3	1910
Nella	4	1940
Rosetta	5	1970
Bella	6	2000
Samantha	7	2030
Angelica	8	2060

FIGURE 2

The eggs in the following two pictures represent four–ounce kittens growing in each mother cat.

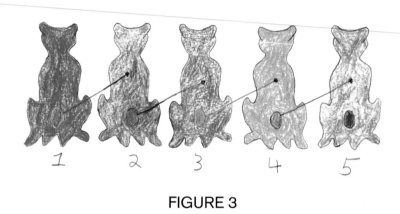

FIGURE 3

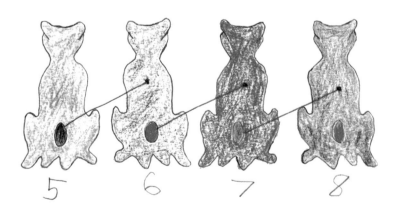

FIGURE 4

As you can see, there is an unbroken line of matter (four ounces) that cannot be in two places at once – an unbroken chain of life. That matter (four ounces) that the kitten Angelica receives from her mother makes it nearly impossible for her to ever go back through time to meet Stephany (her great, great, great, great, great, grandmother) because that would violate the Pauli Exclusion Principle.

The Pauli Exclusion Principle says that the atoms that make up your body can only be in you and cannot be in both you and another object at the same time. The atoms that make up your body at one time started in your pregnant mother, and before they were inside her, those same atoms were inside a star. Let's say you wanted to go back in time to see your great grandmother who passed away when your mother was only five years old. The atoms that make up you can only be in you or in your mother, but not both places at once. If you wanted to see a dinosaur, remember that the atoms that make up your body now also existed back then in the environment. If they didn't you wouldn't exist now. Therefore, the atoms that make up you can only exist in you **OR** in the dinosaur and its environment – but the atoms cannot exist in both places at once.

PART 3

THE PHOTON'S MASS

Physicists have been divided for a long time over how much mass a photon has and whether massless particles can really exist. [25.1, 25.2]

Einstein stated that a photon has mass as per his equation $E = mc^2$. [26.1 – 26.7]

However, other physicists, such as Brian Greene, for some inexplicable reason, claim that the photon has no mass and then say it only appears to have "mass". [27]

They claim that a photon's mass is imaginary, that it must have no mass because it is traveling at the speed of light and therefore has no rest mass. Their reasoning is wrong on many counts. Many people are skeptical about a photon having no mass. For example, Robert Kaplan says that: "some physicists are sure that speaking of a photon's or graviton's mass is meaningful, and that mass is zero – but it isn't anything they can prove." [28]

For example, Sadri Hassani states his BELIEF that a photon has no mass in **Skeptical Inquirer** without any observational and / or experimental evidence to support his belief. [29]

Hideo Nitta, Masafumi Yamamoto, and Keita Takatsu tell us that as a "proof" that a photon has no mass some physicists point to the equation that is used to calculate the mass of an object's motion shows that as the velocity of the photon reaches the speed of light its energy goes to infinity. It is therefore claimed that the only way for a photon to exist without having an infinite mass is if the photon's mass is zero. [30]

However, what they do not tell you about is Godel's incompleteness theorem. There is much debate over what this theorem means but my understanding of it is this: All mathematical axioms cannot be proven true mathematically. An axiom is an abstract mathematical assumption that is assumed true or false. What this means is that there are mathematical statements that are true but cannot be proven true mathematically. [31]

Therefore, there are some mathematical problems in science that mathematics cannot resolve and the mass of a photon is one such problem because we can design experimental tests that show that a photon's energy and mass is non-zero. [32]

There are six other reasons why the physicists who claim a photon has no mass are incorrect. First, just because something is traveling faster than the speed of light or at the speed of light doesn't mean that it has no mass. For example, Miguel Alcubierre came up with a way in which warp drive (which is faster than light travel) might be realized. Let's assume he is correct and that we can build something like the starship **Enterprise**. Does that mean that the starship **Enterprise** loses all mass when it travels at the speed of light or faster? Of course not. [33]

A photon moves in the same way. For example, Einstein imagined what it would be like to ride on a beam of light. He imagined that he was on a light beam looking outward. If you were on a boat traveling on a lake and looked out to the front of the boat as it traveled in the water, you would see the wave that the boat made as it cut through the water. For you, the wave would appear to be static because you and the wave are traveling at the same speed. However, an observer on the shore of the lake would see the wave and the boat moving through the water – both the wave and the boat are in motion. But, relative to you on the boat the wave just sits there – it has no motion. Einstein surmised that if light is like that and you were riding on a beam of light, there would appear to be a wave just sitting there as the beam of light moved through space. [34]

Einstein knew some ways in which a photon could theoretically travel faster than the speed of light. One way in which photons can travel faster than the speed of light is if they travel four–dimensionally. I will explain this in more detail in the second part of the book (see Section 2, Part 2, Proof 4). [35.1, 35.2]

A photon could also theoretically travel faster than the speed of light if it produced its own warp bubble. A photon rides its own spatial wave. Miguel Alcubierre's paper explains everything you need to know about warp bubbles so check it out online. [36]

The second reason that physicists are wrong when they claim the photon has no mass, is that just because an object has no rest mass does not mean it has **NO MASS**. For example, nothing in our universe has rest mass. Think about it – what in our universe has rest mass? For example, if you sit in a chair then you are moving at 1,600 km/hour because, that is how fast the Earth spins on its axis as it orbits the Sun. You are never at rest; you are always in motion because the Earth is always in motion. "Rest mass" only has meaning when two objects are being compared in relation to one another – relative to one another. The spinning of the Earth is what drives wind and ocean currents due to the Coriolis Effect and this is another example of how nothing on Earth has any rest mass. No object in space – planets, moons, comets, asteroids, stars, galaxies – has any rest mass! They all spin around their centers of mass, and at times around each other. Therefore, anything on them is traveling at the same speed they are. For example, if you drive a car at 60 km/hour, you are traveling at the same speed as the car. So, if the car stops suddenly, you keep going, unless you have your seat belt on. The same physics applies with regards to planetary motions. [37.1, 37.3]

In fact, Neil deGrasse Tyson said in at a Larry King interview that if the Earth stopped rotating and you were not seat belt buckled to the Earth then you would fall over and roll about 800 miles an hour due east. [38]

Newton's equation $F = ma$ specified that force equals mass times acceleration. Newton's equation for gravity $F = mg_a$ where force equals mass times gravitational acceleration. [39]

Einstein's equation $E = mc^2$ (which he derived from Newton's equation $F = ma$) energy equals mass times the speed of light squared. [40]

The equation used to calculate momentum is $p = mv$, where momentum (p) is equal to mass (m) times velocity (v). [41]

Momentum is the force (energy) needed to overcome an object's inertia.

Inertia is the innate force of matter's power of resisting change by which matter endeavors to continue its present state, whether it be of rest or of moving forward in a straight line. Inertial mass is equivalent to gravitational mass. For example, the rotation of the planets is due to both inertial mass and their gravitational mass. [42]

The third reason one cannot say photons have no mass is, there are no fewer than five equations in quantum, classical, and cosmological physics stating that any form of energy is some mass propelled at some speed (see Figures 5 and 6).

Energy	=	Mass	Speed
Force	=	Mass	acceleration $V_1 - V_2$
Force	=	Mass	gravitational acceleration
Energy	=	Mass	speed of light squared
Momentum	=	Mass	velocity
Kinetic Energy	=	Mass	velocity^2

FIGURE 5

Acceleration, velocity, and the speed of light squared are all saying pretty much the same thing: how fast an object is traveling, or its speed. All five equations say the same thing, that $E = mv$ (energy is mass times velocity). Energy is a mass traveling at some speed. [43]

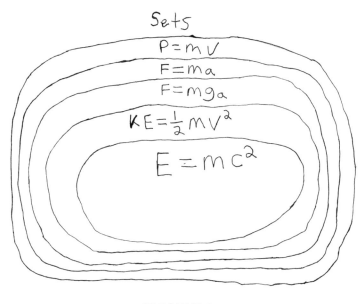

FIGURE 6

James Edward Beichler says that atomic particles (like photons) have a four–dimensional fixed volume. And that saying that a given particle has a constant four-dimensional volume is another way of stating the conservation of matter and energy. For more information see Section 2, Part 2, Proof 3. [44]

Fourth reason one cannot say a photon has no mass is that if a photon can be converted into two particles with known mass then this is a further proof that a photon has mass because something does not come from nothing. [45]

Fifth, Einstein's equation $E = mc^2$ and the other five equations say that any energy particle, like a photon, has mass. No mass equals no energy. Einstein's equation $E = mc^2$ is integrated into the Law of Mass–Energy Conservation. If you divide the energy of the photon by the speed of light you will get its mass. [46]

Sixth reason one cannot say a photon has no mass is that our sun radiates photons, and with each photon that our sun radiates it loses that much mass – in other words each photon carries a little bit of the sun's mass away from it. [47.1 – 47.3]

Brian Greene and others claim that some elementary particles, such as photons, have zero mass; however, Einstein's equation says that photons must have mass and many text books agree with Einstein and they give supporting evidence that backs up their view. [48.1 – 48.5]

Roger Penrose says that an electromagnetic field carries energy, however Einstein's equation of $E = mc^2$ says that it must have mass too. Thus, an electromagnetic field is also matter. This must be accepted because electromagnetic fields are very much involved in the binding forces that hold atomic particles together. There is a substantial contribution (infinite) to any body's mass from the electromagnetic fields in it. [49]

This will make perfect sense when we get to the next part of this section, we will discover that it is impossible to extract all energy from any system and I describe one reason that this is so in Section 2, Part 4 of this section of the book. We will look at how light does work on the universe in Section 2: Part 4. What this means is that light has mass and therefore it exerts pressure, so it is not surprising to discover that light itself can warp space. We will discuss this in the next few parts of this section.

This is connected to the sixth reason that we cannot say the photon has no mass – it is connected to the fact that as our sun radiates electromagnetic radiation (light) it constantly loses mass. As our sun radiates light, the light it radiates creates an outward pressure that counteracts the gravitational pressure that our sun is under. This photonic pressure can be seen in the solar wind – solar sails would not work if the photon had no mass. No mass equals no pressure on the solar sail.

One last thing to note, work is a macroscopic effect on the environment that is done by energy. This is how energy (light for example) does work on our universe.

PART 4

THE CASIMIR EFFECT

One reason that Brian Greene and others would claim that photons have no mass, when physics books use Einstein's $E = mc^2$ equation to prove otherwise, is the Casimir Effect.

There is only one area of physics that says that you can get something from nothing and that is quantum field theory. According to quantum field theory, the Casimir Effect is the result of quantum fluctuations in a vacuum that produce virtual particle pairs (one particle is antimatter and the other particle is ordinary matter), which form out of nothing and then destroy each other, going back to nothing – violating the First Law of Thermodynamics both coming and going. [50]

The problem for modern physics is that the physics that describes very small scales has one set of laws that say that you can get something for nothing. But, at the scale of classical physics there is another set of different laws that say that you **CANNOT** get something for nothing. And then at huge cosmic scales you have a third set of totally different laws that govern things of that scale saying that the universe sprang out of nothing.

When most physicists talk about a Grand Unifying Theory (GUT) to unify physics they only talk about quantum physics and cosmological physics and totally ignore classical physics as if it has nothing to contribute, but it does have a lot to contribute. After all, classical physics gave rise to quantum physics and cosmological physics. For example, even though John C. Hodge does acknowledge the role of classical physics in the origins of cosmological physics and quantum physics he then promptly dismisses the role classical physics plays in modern physics today. After all, we still use many of the laws and rules of classical physics to this very day in many areas of scientific research, like chemistry and engineering for example. [51]

If we had a Grand Unifying Theory (GUT) that would mean, we have one set of laws to govern physics at all scales. That would mean that the set of conservation laws that say you cannot get something for nothing (or from nothing) would work at all scales. Casimir originally intended for the experiment that bears his name to only help him compute the van der Waals force between the polarizable molecules of conducting plates. Thus, the Casimir Effect is not really a proof of zero-point energy of the vacuum – it was never meant to be. [52]

Physicists are divided as to what the Casimir Effect is. First physicists considered the Casimir Effect to be nothing more than the van der Waals force acting on two plates. Today many physicists say that the Casimir Effect is nothing more than some macroscopic thermodynamic effect. [53]

When physicists review NASA's attempt to use the Casimir Effect in some type of rocket or interstellar propulsion device they are highly critical and skeptical of the approach NASA is taking. [54.1, 54.2]

However, quantum theorists think that the Casimir Effect is proof that quantum fluctuations in a vacuum produce virtual particles that somehow affect the plates of the Casimir Effect Experiment. Quantum theorists have yet to prove the existence of such particles and they have yet to measure their size when they come into existence, also quantum theorists have not measured the energy level of the particles when they come into existence then wink out of existence. All these quantum fluctuations in a vacuum, sounds too pipe dreamy to many physicists.

For example, Roger Penrose states that, "the experimentally established existence of the Casimir Effect does *not* actually establish the physical reality of vacuum energy." [55]

The physics community is divided as to whether the Casimir Effect is the result of the van der Waals force or the result of quantum fluctuations of virtual particles in a vacuum. There is an experiment that can settle the issue once and for all. But before we get to that let us look at why there is so much controversy regarding the Casimir Effect.

One last point that must be mentioned is antimatter. Antimatter is what is supposedly produced in a quantum vacuum and it is also supposedly produced when cosmic rays from the sun strike Earth's atmosphere. Many of the problems that are associated with supposed quantum fluctuations in a vacuum are also associated with antimatter theory. Even if antimatter does exist, there are questions as to exactly **HOW** it interacts with "normal" matter. For example, antimatter particles have been seen to act normally enough in cloud chambers leaving vapor trails. The antimatter should instead detonate when the antimatter particle goes through the vapor and hits the vapor molecules as it passes through the cloud chamber. But the antimatter does not do this. What might be going on here? [56]

The issue here is the velocity to which we subject particles – matter and antimatter – in particle accelerators. If I have two pieces of poo, and leave one on the floor, and sling the other at the speed of light at a brick wall, the resulting explosion would not be the manifestation of the nature of the poo, but it would instead be the result of its velocity when it hit the wall. Remember, we have five equations that tell us that energy is a given mass traveling at a given velocity (speed). Antimatter reacts in the same way.

If you gently place a lump of antimatter on the floor, it most likely will do nothing but sit there. Now take another lump of antimatter and throw it at a wall at the speed of light and the results will be quite different because of the velocity at which the antimatter was traveling.

Another point that must be mentioned but will be covered in the second half of this book is Hawking Radiation, which is related to quantum fluctuations in a vacuum.

Quantum theorists want the Casimir Effect to be due to quantum fluctuations of virtual particles, because they believe that this would help them understand how the universe began and it would help them to understand the universe's current accelerated rate of expansion. What the quantum theorists do not understand is that these issues, and many more, can be better understood using classical physics – you cannot get something for nothing (or from nothing) – to explain the state of our universe's origin and its current rate of expansion.

For example, let us see how the laws of thermodynamics can be used to understand the expansion of the universe. Brian Greene brought up the question of 'what the temperature of the space that is surrounding our universe might be'. He hinted that it was hotter than the universe itself. Greene thinks that the theory of inflation works like this. He tells us to imagine that a block of cheese is all of space before the formation of stars and galaxies. According to the idea of inflation, space (the cheese) is saturated with a huge amount of energy. It is this energy that causes space to expand at an enormous speed. As it expands, energy randomly discharges like a spark of static electricity, but this is static electricity that is on a cosmic scale. When this cosmic static electricity discharges, all that energy is converted into particles of matter. This process gives birth to a new universe, a Big Bang. These new universes are like holes in the energy-saturated space (the cheese). Inside of each universe, space continues to expand but at a much slower rate and sometimes stars, planets, and galaxies form in these universes. [57]

So, what he is describing here seems to an infinite amount of space that is like hot lava in which big bangs of energy occur, and this somehow produces ice-cube universes (our universe is only $3°$ Kelvin). Yeah, I don't think so because we do not see this in nature and classical physics. Also, if this were the case then the universe would be shrinking instead of expanding, because according to the Second Law of Thermodynamics, heat travels from hot to cold bodies. Therefore, if the space outside of our universe were hotter than our universe we would see and feel it because our universe would be rapidly shrinking, collapsing in on itself in a Big Crunch. [58]

The following figures show this. Figure 7 is the Key to Figures 8 – 12. Figures 8 – 9 are showing the temperature outside of our universe as being hotter than $3°$ K absolute zero. Figure 10 shows the temperature outside of our universe as being the same temperature as our universe at $3°$ K. Figures 11 – 12 show the temperature outside of our universe as being colder than $3°$ K above Absolute Zero.

Temperature

	10,000,000,000,000,000,000 Kelvin
	10,000,000,000,000,000 Kelvin
	10,000,000,000,000 Kelvin
	10,000,000,000 Kelvin
	10,000,000 Kelvin
	10,000 Kelvin
	10 Kelvin
	9 to 8 Kelvin
	7 to 6 Kelvin
	5 to 4 Kelvin
	3 to 2 Kelvin
	1 to 0 Kelvin
	-10 Kelvin
	-10,000 Kelvin
	-10,000,000 Kelvin
	-10,000,000,000 Kelvin
	-10,000,000,000,000 Kelvin
	-10,000,000,000,000,000 Kelvin
	-10,000,000,000,000,000,000 Kelvin

FIGURE 7

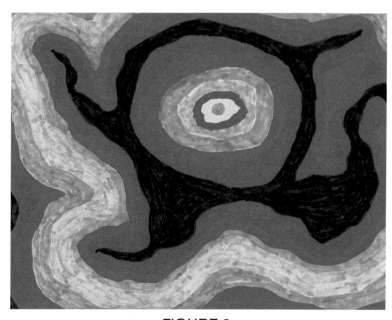

FIGURE 8

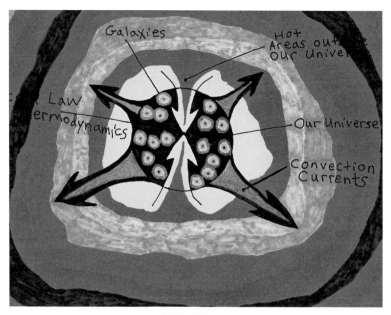

FIGURE 9

If the temperature of our universe and the space outside of our universe were the same, then we would have a static universe – a steady state. Something that Einstein thought until Hubble's observations of the motions of galaxies proved otherwise (see Figure 10).

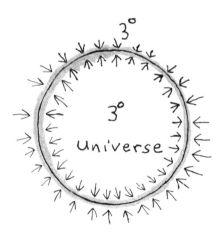

FIGURE 10

But if the space outside of our universe is colder than three degrees above Absolute Zero then we would have an expanding universe that is obeying the laws of thermodynamics – no dark matter, dark energy, or quantum fluctuations in a vacuum needed. The rate of expansion can be put into a mathematical formula to calculate the temperature that is needed in the space that surrounds our universe to achieve the observed expansion rate of our universe. [59]

Our universe shines like a star compared to the -10,000° Kelvin material that our universe is surrounded by. But wait – don't most scientists agree that you cannot get below 0° Kelvin because once you are at 0° Kelvin, you have extracted all the energy from the object or system?

There are three reasons that you can get temperatures billions of degrees minus Kelvin. The first reason being that Mother Nature does not have to bow to the whimsy of human rules, but that aside, the second reason is the First Law of Thermodynamics comes into play here – if you extract all the energy from an object or system, it is the same as destroying all energy in the object or system, and it is impossible according to the laws we know. The third reason gives us a mechanism by which all the energy of an object or system can never be extracted. For example, if as Carl Sagan suggested, that we do indeed live in a Fractal Universe, then it would be impossible to extract all the energy from an object or system because each object or system has an infinite amount of energy within it.

For example, if we use the following analogy – that our universe is like a volcano. If we lived in a volcano it would make sense to say that temperatures of zero degrees Celsius (the freezing point of water) is impossible. However, it would be incorrect to say that temperatures of zero degrees Celsius are impossible outside of the volcano. Temperatures below 'Absolute Zero' maybe possible outside of our universe in a similar way.

FIGURE 11

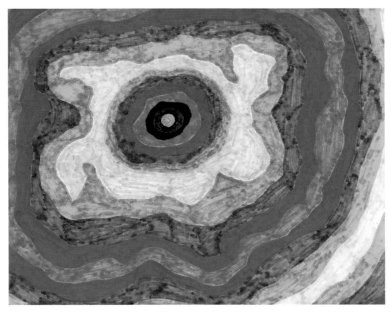

FIGURE 12

It was discovered that huge cold Cosmic Voids in our universe are exerting a repulsive force that helps to expand our universe. Then there are huge clumps of intensely hot dense matter are attracting more matter to themselves forming huge Filaments full of galaxies. Cold voids expand space and hot areas of the cosmos contract space, thus drawing more mass toward them. [60]

The Comic Voids have low pressure, which results in low temperature. Cosmic Voids act like mountains or hills. Istvan Szapudi describes how Cosmic Super Voids act like mountains he thinks that the reason these voids act like hills or mountains is that there is very little matter in these voids, which also means there is very little energy. So, any photon of light that enters these super cold regions of space lose energy as if it were rolling up a hill. But when the same photon leaves the void – in a static universe the photon would arrive on the other side of the void with the same energy it had when it entered. However, the accelerated expansion of the universe changes things. In an expanding universe, when a photon enters the void, it loses energy, but when it begins to leave the void, it cannot regain all the energy it had when it entered, the void because the void is expanding along with the rest of the universe. [61]

In Figure 13 we see the path of a photon as it travels across a cosmic super void in a **static** universe. In Figure 14 we see the path of a photon in an **expanding** universe as it crosses a Cosmic Super Void.

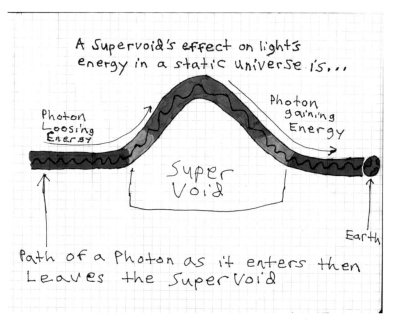

FIGURE 13

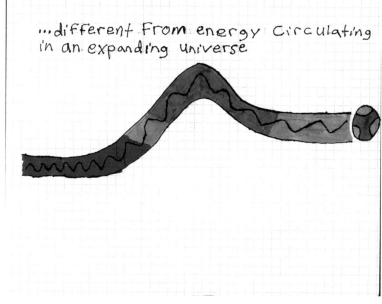

FIGURE 14

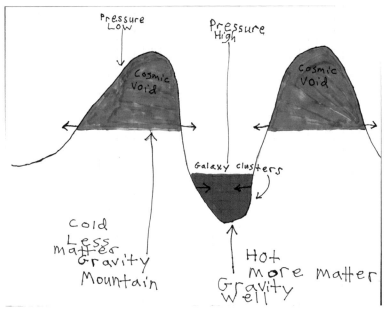

FIGURE 15

There are two questions I would like to address. The first question is: if at 0° Kelvin you have extracted **ALL** the energy from an object or system then wouldn't you be unable to get temperatures below 0° Kelvin? There are five reasons why we would be able to get temperatures below 0° Kelvin and the first reason is the First Law of Thermodynamics which states that energy cannot be made nor destroyed – removing all energy from an object or system is the same as destroying all energy from an object or system, and therefore, this would be impossible – you cannot get something from nothing.

Second, nature does not have to obey our rules, on the contrary it is we that must obey nature's rules.

Third, didn't we discuss the Fractal Cosmos earlier? If Carl Sagan and the other scientists that hold with such a possibility are correct, then this would provide a very powerful mechanism to uphold the First Law of Thermodynamics by making it impossible to extract all the energy from any object or system.

Fourth, there are five states of matter – plasma, gas, liquid, solid, and Bose – Einstein Condensate. Humans have made Hydrogen into a plasma, gas, and liquid, but it has not been made into a solid. This implies that even colder temperature than 0° Kelvin are possible.

Fifth, there was always some doubt as to whether you really could extract all the kinetic energy from any system. For example, an electron microscope will show atoms in a solid jostling about, so not even when you "freeze" something solid can you extract all the energy from any system. Robert E. Krebs states that some people think that at Absolute Zero all molecular motion ceases at this temperature. Robert E. Krebs also says: "This is not quite correct since molecules of solids continue to 'vibrate' but not move at random or exhibit any kinetic energy as molecules would in matter at higher temperatures." [62]

The second question I would like to raise is – what is outside of our universe that is billions of degrees below 0° Kelvin? What could possibly be that cold? If you look at the periodic table of elements, you will notice that all the elements on the periodic table, except hydrogen, are made by stars or the explosion of stars. But there is no known way to make hydrogen. Stars start off with hydrogen, but they do not make it (see Figure 16).

In Jennifer Johnson's article in **American Scientist**, there is a periodic table that is like the one in this book. In her periodic table ONLY HYDROGEN came from the Big Bang. All other elements came from other processes at work like fusion, fission, and radioactive decay. [63]

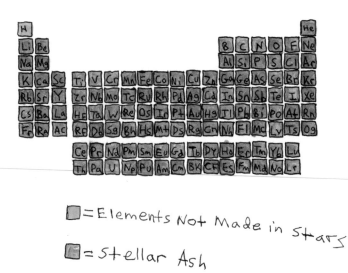

FIGURE 16

If it is true that our universe resides in an infinite expanse – a sea of icebergs – of Bose – Einstein Condensate Hydrogen at temperatures around -10,000° Kelvin then the expansion rate of our universe (303,630.936 km/s) would press against the Bose – Einstein Condensate thus producing a ram pressure against our universe as our universe moves through it at 303,630.936 km/s. This ram pressure would flatten any mountains and hills of the Super Voids (see Figures 15 and 17). [64.1 – 64.3]

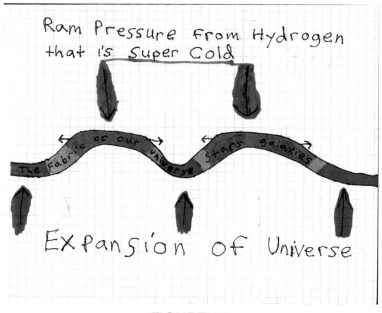

FIGURE 17

Another interesting question is: What is the nature of the Bose – Einstein Condensate Hydrogen that surrounds our universe? Could it be acting like a super conducting fluid with ice burgs of Bose – Einstein Condensate Hydrogen distributed through it? John Langone, Bruce Stutz, and Andrea Gianopoulos state that in the 1950's there were three American physicists who argued that when atoms are in a super–cold state, they arrange themselves into distinct geometrical arrays. When the atoms do this, their electrons form into pairs that emit and absorb energy equally, so there is nothing to impair their movements. For example, the atoms cooled to 2.19^0 Kelvin all have the same momentum. These atoms are like running horses tied together; if one moves, they all move. Any heat that is produced in these super–cooled atoms is conducted so quickly that it forms a wave through the material – that is why they are called superconductors. When a magnetic field approaches a superconductor, it causes swirling electromagnetic currents in the outermost layers of these super conducting materials that shove all other electromagnetic fields away. All superconducting materials will levitate above any external magnetic fields. [65]

As our universe rotates around its center of mass, does it create a super conducting magnetic field that repels other universes? The possibility of rotating universes was first proposed by Kurt Gödel. Neutron stars have a superfludic core that creates a huge magnetic field. Our universe might just be swiftly rotating in the sea of Bose – Einstein Condensate Hydrogen that surrounds our universe. [66.1 – 66.5]

If, as Carl Sagan surmised, our universe is just an elementary particle in a much grander universe (a fractal universe), what then does this say about the universe or multiverse as a whole? Is it possible that our universe is like a superconductor that is suspended within a superfluid sea of Bose – Einstein Condensate Hydrogen with zero viscosity?

Astronomers have long assumed that the mass of our universe was critical to understanding how it evolves over time. They also used the mass of the universe to predict how the universe expands and its rate of expansion. However, we have looked at a new way of interpreting the data. Mass is still critical to unlocking the expansion rate of the universe but now when it is coupled with the mass's temperature, we get a much clearer picture of how and why the expansion rate is the way that it is.

Richard L. Faber says: "Einstein proposed that gravity is not a force but a curvature of spacetime!" [67]

Steven Holzner says that Einstein suggested that space and time are different aspects of something called 'spacetime' with mass and energy able to curve the 'spacetime' – it is this curvature that we call gravity. [68]

We have shown that high concentrations of hot matter warp space by contracting it and high concentrations of cold matter warp space by expanding and stretching it also.

Astronomers have long assumed that the mass of any object – our universe included – determines its gravity (that is, how it warps the space around it). But now we understand that mass alone does not determine the effects of gravity (the warping of space); it is the combination of mass and temperature that determines the warping of space (gravity).

Alex Filippenko in a speaking engagement at **Google** described how at first our universe decelerated, and then after a while it began accelerating. He states that the reason for that is that when our universe began its expansion galaxies were closer together, and this means their gravitational attraction for each other was stronger than it is now. When our universe started its expansion, any repulsion in it if it was a property of space or something within space was relatively minor. As our universe expanded the gravitational attraction between the galaxies declined over time. However, the repulsion of space itself increased with time. [69]

Alex Filippenko said: "If the repulsion is caused by something in space or a property of space itself, then the more space there is, the greater is the repulsive effect." That means the force of gravity is decreasing while the force of antigravity is increasing. These forces cross at some point. The point where they cross is when the universe starts accelerating in its expansion rate. The real prediction for Alex Filippenko, Adam Riess, and others was that if they could look back far enough in time, they would see a phase at which the universe was decelerating. Filippenko and Riess were part of a Hubble project that looked for and monitored very distant supernovae that are 7, 8, 9 or even 10 billion light years away. What they found was that the data that they gathered confirmed their prediction of the early deceleration of the universe. Their data showed that the universe was steadily decelerating until about 5 billion years ago when it started to accelerate. They had discovered that our universe's expansion rate had changed from deceleration to acceleration. Mathematically, when you go from deceleration to acceleration that change is known as a jerk. A jerk is a non-zero third derivative of position, you have position, velocity, acceleration, jerk. [70]

We can interpret this data in terms of thermodynamics as seen in Figures 18 – 20.

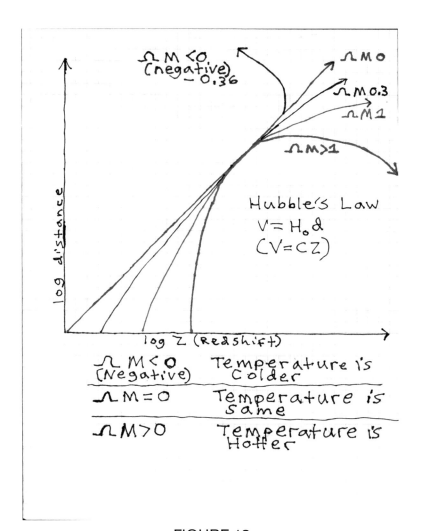

FIGURE 18

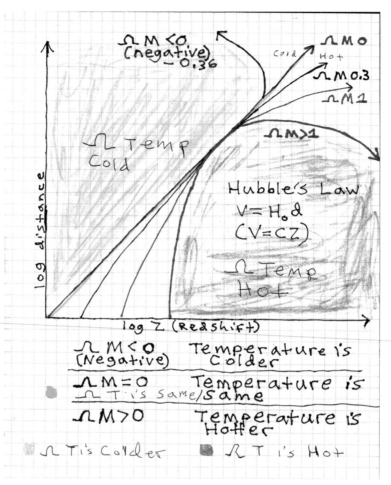

FIGURE 19

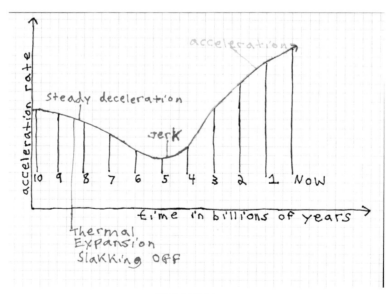

FIGURE 20

Let us now turn to the Casimir Effect. According to quantum field theory, a vacuum has quantum electromagnetic fluctuations that can be detected if two metal plates are placed in the vacuum. If the two metal plates are separated, then they will slowly come together because the quantum electromagnetic wave fluctuations push the plates together. This is like gravity (which does the same thing), so what is the difference between the Casimir Effect and gravity? There may be no difference at all – the Casimir Effect has nothing to do with electromagnetic waves but instead it has to do with the heating and cooling of space itself, which is indistinguishable from gravity. Experiments have shown that the Casimir Effect is thermally driven. [71]

Quantum physicists have not definitively proven that the Casimir Effect is caused by quantum electromagnetic fluctuations in a vacuum. First, the quantum physicists have not proven that any electromagnetic field exists by detecting it. Second, the quantum physicists assume that quantum particle pairs spontaneously form and annihilate each other but the energy that should be released in such an event has never been detected either. This spontaneously coming into existence and then winking out of existence is a violation of the First Law of Thermodynamics – coming and going. It is highly doubtful that quantum fluctuations in a vacuum exist at all.

Quantum physicist totally ignore Casimir's work. In Casimir's original paper he never mentioned quantum fluctuations in a vacuum. He did mention that he was studying the van der Waals effect and classical electrodynamics. He did use the term "zero-point energy", but he says that: "This force may be interpreted as a zero-point pressure of electromagnetic waves". [72]

Zero-point energy in Casimir's Experiment means that no photons (or any other radiation) is originating from the gap between the two conducting plates. Therefore, there is no radiation that is striking the conducting plate's surface to provide the pressure needed to keep the plates apart. [73.1 – 73.2]

Steven Holzner says that when the matter and antiparticle that pop into existence and then wink out of existence in a quantum vacuum, they should produce gamma waves which you would think would be quickly detected **IF** the quantum vacuum did produce 'virtual particles.' [74]

Many scientists think that the Casimir Effect is really a macro van der Waals Effect. [75]

However, other scientists are still looking for reasons to believe in these quantum fluctuations in a vacuum, so they develop new tests. One of these tests involves hydrogen atoms. The researchers claim that the hydrogen atoms are jiggled about by virtual particle pairs as they come into existence then annihilating themselves. But, here again there is zero evidence that there are virtual particles popping into and out of existence. These 'virtual particles' have never been seen. No one has ever witnessed them coming into existence then going out of existence – no photographic evidence has ever been presented. When these 'virtual particles' come into existence there should be a large energy spike, alas this has never been witnessed either. When these 'virtual particles' wink out of existence there should be an even bigger energy spike as the matter and antimatter 'virtual particles' meet to annihilate themselves, and here again these energy readings were never recorded. If the hydrogen atoms did jiggle, then it could have been from almost anything. Atoms, especially gasses always jiggle because everything in our universe is in motion. The hydrogen atoms could have jiggled due to thermodynamics, a cosmic ray, or Brownian Motion. [76]

This means that the mechanism that truly causes the Casimir Effect is to this date largely unknown. If indeed the Casimir Effect and gravity are thermally driven, then they may well the same thing. This would also mean that on a quantum or cosmic scale gravity is thermal pressure fluctuations that cause wave propagation (gravity waves). If this is the case, then mass, temperature, and pressure are all proportional to each other. We see this in stars, as their mass increases so does the temperature and pressure.

Adam Riess and Michael Turner states: "An object's gravity is proportional to its energy density plus three times the pressure." A good example of this is our sun. Our sun is a sphere of hot gas. Since our sun is composed of hot gas that has a positive [outward] pressure that counter acts gravity's negative [inward] pressure it will remain shining for billions of years. However, gas pressure rises with temperature and that means that the sun's gravitational pull is slightly greater than a cold ball of matter that is equal to our sun's mass. However, a gas of photons has a pressure that is equal to one-third its energy density; therefore, its gravitation will be twice that of an equal mass of cold matter. [77]

Riess and Turner realize that there is something not quite right with quantum mechanics' notion that the Heisenberg's Uncertainty Principle requires that a vacuum be filled with particles living on borrowed time and energy, popping in and out of existence. The problem is that when quantum theorists compute the energy density of the quantum vacuum their numbers are incorrect. The quantum theorists come up with values for the quantum vacuum that are 55 orders of magnitude too large. If the energy in the vacuum were really that high, then all matter in the universe would instantly fly apart. [78]

Adam Riess and Michael Turner recognize that this discrepancy between observations and the calculation of quantum physicists is one of the worst embarrassments in all theoretical physics. Riess and Turner also think that this may be a sign of a great opportunity. Even though they think it may be possible that some new attempts to establish the vacuum energy density may yield the right amount to explain cosmic acceleration, Riess and Turner think that calculations incorporating a new type of symmetry will lead physicists to the conclusion that any energy associated with the quantum vacuum is zero. According to Riess and Turner, even quantum nothingness weighs nothing. Remember we already established that all energy has mass ($E=mc^2$); therefore, if a vacuum has a lot of energy then it would also have a lot of mass. Riess and Turner feel that something else must be causing the expansion of the universe to accelerate. [79]

Many physicists feel that whatever drives the Casimir Effect is very likely responsible for the expansion of the universe. If the Casimir Effect is thermodynamic, then we would have a very good idea as to what causes the expansion of the universe. There is an experiment that we can run that will prove that the Casimir Effect is thermodynamically driven, with no quantum fluctuations in a vacuum needed.

ROGER I. PARKER II

PART 5

THE MODIFIED CASIMIR EFFECT EXPERIMENT

For example, we just showed that Classical Thermodynamics can be used to accurately explain the expansion of the universe and this fits in with the observations that astronomers have made over the years. Now we have a more realistic way to create a warp bubble around a spacecraft. To create a warp bubble around a spacecraft you heat up the space in front of it thus warping (contracting) the space in front of the spacecraft, then you want to cool the space behind the spacecraft to warp (expand) space behind the spacecraft (See Section 3, Part 1). Now to prove that space can be warped by thermodynamic influences then we need a specially designed test. So, let us run a Casimir Effect experiment that can pin down exactly what is causing the plates to move. To show that the Casimir Effect is in fact a macroscopic thermodynamic event, we will build a vacuum chamber in the following way.

This Modified Casimir Effect Experiment is designed to visually measure the thermal fluctuations in a vacuum. First, we build a ceramic vacuum chamber with infrared cameras embedded in the walls that will allow us to look at what is happening in the vacuum chamber and visually show and measure all the heat within the vacuum chamber. The vacuum chamber will also have induction-heating units embedded in the walls. The parallel plates need to be hung from a gypsum rod because gypsum has poor heat conducting properties. Which means that little to no heat gets transferred to the parallel plates through the rod. Any heat transfer from the rod to the plates would interfere with the outcome of the experiment. To further ensure that heat transfer from the rod to the plates does not occur we could add thermal sensors to the rod and employ a device that automatically heats or cools the rod to keep it the same temperature throughout the experiment – keeping the rod thermally neutral. The last thing we need to do is place the whole modified vacuum chamber within a tank that can be filled with liquid nitrogen (see Figures 21 and 22).

Key

1 Modified gypsum rod and parallel Plates

2 Ceramic vacuum Chamber with induction Heating units

3 Infrared Thermal Imaging Cameras

4 Liquid Nitrogen Tank

FIGURE 21

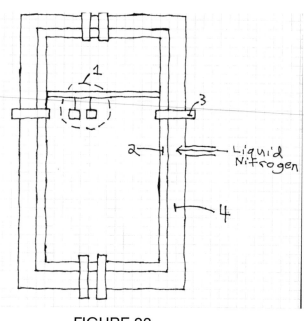

FIGURE 22

You may be asking; "Why do we need to make all these changes?" To answer this question and try to pin down exactly what is causing the plates to move, we need to run the experiment then you will see why we needed to make the changes. So, let us run the Modified Casimir Effect Experiment and try to pin down what exactly is causing the plates to move.

When we conduct the first part of the experiment, we will heat up the vacuum chamber. As the walls of the vacuum chamber heat up some of that heat is radiated, in the form of infrared photons, to the vacuum and then to the parallel plates. The infrared cameras will record this, and they will also show the parallel plates being pushed together by the infrared thermal radiation (see Figures 23 and 24).

In the second part of the experiment, we will quickly cool the vacuum chamber by pumping in liquid nitrogen into the tank that surrounds the vacuum chamber. As the walls of the vacuum chamber cool down the hot parallel plates will radiate thermal radiation, which will slowly pull the plates apart as they cool. The infrared cameras will record the actions of the plates as they move apart.

The thermal radiation from our Sun travels through the vacuum of space to reach us on Earth; just like we expect the thermal radiation of the walls of the vacuum chamber to reach the parallel plates to push them together.

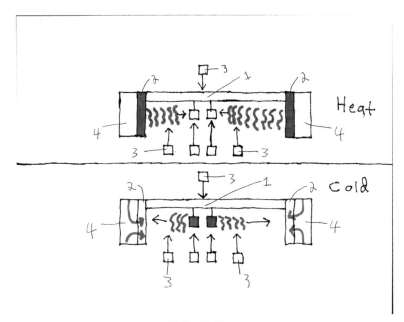

FIGURE 23

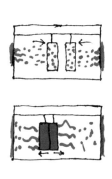

FIGURE 24

PART 6

CONCLUSION

In this way, we can infer that space reacts by warping as a result of thermal pressure. Einstein wanted to unify gravitation with electromagnetism. All forms of light – including infrared (thermal radiation) are electromagnetic. This is a consequence of the Photovoltaic Effect. [80]

What this experiment shows is that electromagnetic – photovoltaic thermal radiation warps space (remember, gravity's definition is the warping of space – more specifically the four–dimensional warping of space). This gives us a cause – effect relationship to unite the Weak–Electromagnetic–Photovoltaic Force with the gravitational (the warping of space) effect. In this way we can fully understand why gravity is not a Force it is an effect of a Force acting on space itself. In this way, the Grand Unification that Einstein longed for will be finally achieved. Not only that, this knowledge can lead to a better and more realistic theory of quantum gravity. All of physics is ruled by thermodynamics because light does work on the cosmos at every level. Quantum Physics is ruled by thermodynamics, Classical Physics is ruled by thermodynamics, Cosmological Physics is ruled by thermodynamics just get over it.

1. We deduced that what is outside of our universe must be colder than our universe's current temperature.

2. We observed and recorded the Thermal Effect in the cosmos – Cosmic Voids and Cosmic Hot Spots.

3. We have proof that our universe is expanding and that our universe is experiencing an increase in the rate of its expansion.

4. We deduced that there is a vast ocean of super cold hydrogen outside of our universe.

5. We have experimental evidence that heat can affect what happens in a vacuum and how the vacuum reacts to temperatures. This experiment proves that gravity (the warping of space) is driven by not just mass, but mass and its temperature. In other words, gravity (the warping of space) is caused by thermal pressures – the greater the temperature the greater the pressure exerted on a given space and hence the greater the warping of space (gravity). In conclusion – gravity is the effect of thermal radiation pressure. Einstein tried and failed to unify electromagnetic – photovoltaic radiation with gravity. We have experimentally shown a clear cause (thermal – light) effect gravity (warped space) relationship between the two.

There is more in the Second Section of this book and in this section, we will explore the fourth dimension and look at the four–dimensional warping of space in more detail.

Another issue is brought up by Bill Bryson. He says that; it is said that atoms themselves contain empty space. Bill Bryson says that an atom's nucleus is only one millionth of the full volume of the atom. An atom's nucleus is also very dense, because it contains virtually all the atom's mass. By comparison if an atom were the size of a cathedral, like the Saint Peter's Basilica at the Vatican for instance, then the nucleus would be the size of a fly at the center of the cathedral – but this fly sized nucleus would be many thousands of times heavier than the cathedral. [81.1, 81.2]

Bill Bryson then goes on to say: "It is still a fairly astounding notion to consider that atoms are mostly empty space." [82.1 – 82.2]

What this means is that if we make an atom the size of Saint Peter's Basilica then the 5,000,000 m^3 gross volume of the cathedral would be empty space with a fly sized nucleus at the center. If we believe that empty space produces virtual particles out of nothing (violating the First Law of Thermodynamics in the process), then wouldn't these virtual particles as they pop into and out of existence in the empty space of the atom disrupt the atomic nucleus itself and all the electrons flying around it. After all, Hawking Radiation and the Casimir Effect are attributed to fluctuations of virtual particles in a vacuum. This is shown in Figure 25.

John D. Barrow says that if an electron is placed in a quantum vacuum the electron will attract the oppositely charged particle from the many virtual particles that briefly pop into and out of existence from the quantum fluctuations in the vacuum. [83]

If this is true, then that means that any virtual particle pair that pop into existence in the quantum vacuum that exists in the atom will begin to interact with the electrons that surround the atomic nucleus. These interactions will occur in the following way. First when a pair of virtual particles pop into existence near an electron as it orbits around the atomic nucleus, the positively charged virtual particle will be attracted by the electron. When this happens the negatively charged virtual particle will fly off – perhaps it will be attracted by a proton of the atomic nucleus. There are two different outcomes for all this.

The first outcome would be, if each virtual particle were destined to annihilate itself with its oppositely charged partner then in the case of a hydrogen atom, for example, once the electron attracts the positive charged virtual particle to it and the proton nucleus attracts the negative charged virtual particle to it then the whole hydrogen atom should explode thus annihilating it. Atoms do exist in our universe so that means that this scenario is highly unlikely.

The second outcome is championed by Stephen Hawking. He said what if these virtual particles don't have to annihilate each other then you could detect these virtual particles coming from 'black holes' because one of the virtual particles would enter the 'black hole' and the other particle would be detected as radiation (Hawking Radiation to be exact). In this scenario what you would have is the virtual particle pair would pop into existence in the vacuum inside the atom then the positive charged virtual particle would be attracted to the electron and the negative charged virtual particle would be attracted by a proton of the atomic nucleus. If this were to happen to a hydrogen atom, then the proton and the negative charged virtual particle would pair up and possibly form a new atom. What happens to the positive virtual particle and the electron depends on a lot of assumptions. The first possible outcome would be for the positive virtual particle (if it was a positron) to combine with the electron and either annihilate each other or they could create a low mass neutron. At best this scenario is very improbable and at worst it is impossible. Plus, we have experimental evidence (see Section 1 Part 5) showing that quantum fluctuations in a vacuum do not exist.

Also, it is very unclear what differentiates a 'quantum vacuum' from any other vacuum that might exist. Are all vacuums 'quantum vacuums'? If not, then how do we distinguish one from the other?

Quantum fluctuations in a vacuum would be a perfect mechanism by which the magic spells that wizards and witches could cast would function – the physics of magic. For example, quantum fluctuations in a vacuum could be made to produce fire from a mage's fingers much like Hawking Radiation. Chairs could pop into existence out of thin air. However, we do not see these things therefore, it is highly doubtful there are any quantum fluctuations in a vacuum that ever could produce virtual particles. The whole idea seems to be a bunch of magical thinking. There are some scientists, like Amit Goswami, that believe that human consciousness creates the material word around us. This is the philosophy of a child. Children believe that their thoughts, wishes, and desires have the power to affect reality. [84]

The Anthropic Principle is a childish notion that the universe is the way that it is because if the universe were different then we would not be here to observe it. [85.1, 85.2]

The Anthropic Principle takes many forms. For example, our universe is not Anthropomorphic; it is not made in the image of humans. Our universe is not Anthropocentric; our universe does not revolve around humans or human needs. Our universe is not Anthropometric; humans are not the measure of our universe. Our universe does not depend upon human theories, beliefs, observations, or perceptions. Our universe exists independently from human observations. [86]

How we perceive our universe evolves over time. When new information proves that our old views (perceptions) of our universe are wrong we invent new perceptions to fit the new information. Our view of the universe constantly changes over time.

To conclude this Section of the book we must look at abstract mathematics and how that influences all the magical thinking that we observe in science today. For example, Matt Parker says Dirac used abstract mathematics to predict the existence of the anti – electron. This led Dirac to believe that there might now be a new way of doing physics. Dirac thought that theoretical physicists needn't worry about doing experiments any more. Instead, the theoretical physicists should play around with mathematical theories. Today if you have beautiful math supporting an otherwise untested theory this is a kind of 'tick of approval' for some physicists, because they think that the universe likes elegant math regardless of how abstract it may be. Matt Parker hopes that such mathematical theories might one day be put to the test in experiments. [87]

However, Roger Penrose is among the skeptics of quantum mechanics and this is what he said in an interview by **Discover** magazine was that Schrodinger, Einstein, and Paul Dirac were all skeptical of quantum mechanics. According to Penrose, people are surprised to find out that Dirac was skeptical of quantum mechanics because he laid the foundation of quantum mechanics. Many people think that Dirac was a defender of the quantum mechanics he helped to set up, but truth be told that is not the case. For example, when someone asked Dirac, 'What's the answer to the measurement problem in quantum mechanics?' his reply was, 'Quantum mechanics is a provisional theory. Why should I look for an answer in quantum mechanics?' Dirac did not have full confidence that quantum mechanics was true. Must people don't know this because Dirac didn't say this out loud much. [88]

The modern equivalent of quantum mechanics is string theory. Matt Parker says that in some cases, String Theory, requires 11–dimensional (11D) spacetime. However, Einstein's theory of relativity only requires 4D spacetime. While we have a lot of evidence supporting Einstein's theory of relativity, we have zero observational evidence that supports String Theory. Even though string theory is just a mathematical theory, Matt Parker thinks it would be such a waste if the beautiful math behind String Theory were not true. On the other hand, just because we humans think something is beautiful does not mean the universe agrees. [89]

The abstract mathematics of quantum mechanics and string theory has no underpinnings in reality. Measurements cannot be made – it's like asking an engineer to build a stair well from an abstract painting of a stair well. The engineer would think you were nuts because the abstract stair well has no underpinnings in reality – there is nothing for the engineer to work with because what can the engineer draw measurements from. Quantum mechanics and string theory's cult like reliance upon abstract mathematics is a detriment to physics and science.

According to James Edward Biechlier the math that is used in quantum physics became more abstract and meaningless when it was "rigorized" by purging all the math that is used in physics of all references to the physical world we see around us every day. By doing this many quantum theorists feel that they 'purified' the mathematics that they use by making it more abstract. In reality, however, they make the mathematics that they use more and more useless the more that they "rigorize" it. And when the quantum physicists apply their abstract math to any event in the physical world, they find that their math cannot adequately describe the physical world any more so they start to make up more and more things to compensate for their total lack of understanding. The 'quantum fluctuations in a vacuum' is one such example of quantum physicists making something up to hide the fact that the mathematics that they use no longer has any relevance or meaning. [90]

In fact, math is NOT a science. As Robert Scherrer says, "…mathematics does not follow the experiment / theory model and is therefore not a science, despite being indispensable to scientific progress." [91]

Mathematics should never be a replacement for rigorous scientific experiments. Math is nothing more than a tool of science. Math is a tool that can be misused.

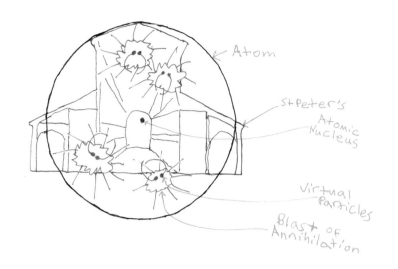

Atom

St.Peter's

Atomic
Nucleus

Virtual
Particles

Blast of
Annihilation

FIGURE 25

THE FOURTH DIMENSION

PART 1

THE FATE OF THE UNIVERSE

In the first section of this book, I set the stage for the material that I will present here. We will look at the four–dimensional nature of our universe. But first, let us tie up a loose end.

The universe, as you have seen in the preceding section, is doing more than just expanding from or contracting to a central point. Voids push matter while huge galactic clusters pull matter. The space in our universe is fluidic in the sense that temperature variations cause matter to travel from one place to another like a fluid.

The old models of our universe's fate do not fit into this new dynamic model of the universe. With the previous models only, mass determined what gravity would do. That meant that only two options were possible. The first option was that as the universe expands away from a central point the galaxies would move away from one another and the voids between them would grow larger. Until there are no other galaxies in the Observable Universe. At which point the universe would become a cold haze as all the stars in our galaxy start to wink out. This view of the future of the Observable Universe is seen in Figure 27. We have not observed this. In fact, we have observed the opposite. We have seen thermal currents pushing galaxies out of voids and into huge galactic clusters. That means this theory must be abandoned. [92]

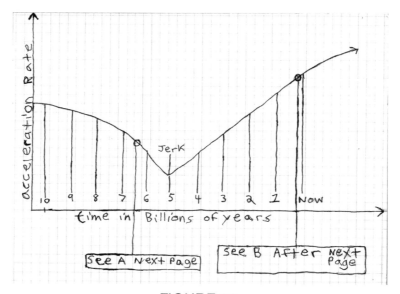

FIGURE 26

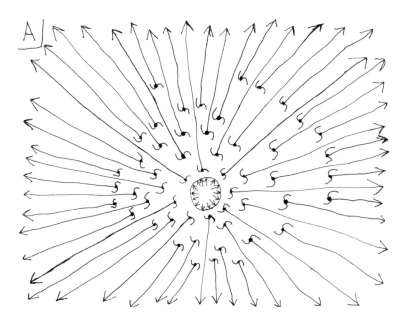

FIGURE 27

The second option stated that the universe would stop expanding and all the galaxies would start moving toward one another to a central point – a Big Crunch. This view of the future of our Observable Universe is seen in Figure 28. This theory must be abandoned also. We have observed galaxies collecting together in huge galactic clusters, but these clusters do not converge on a central point; instead, they form massive filaments. Our Observable Universe resembles a huge sponge and the two previous models of the development and fate of our universe do not take the sponge like structure of our universe into account. [93]

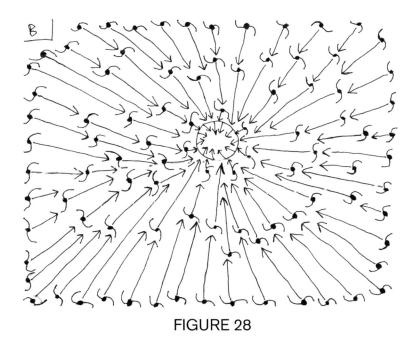

FIGURE 28

There are many theoretical models of the universe that cosmologists have devised, but only one of these theories fits all the observational data that we have so far. [94.1, 94.2]

This theoretical model of the universe is called the Friedmann – Lemaitre Universe. [95]

The Friedmann – Lemaitre Universe model shows the development of a universe that begins with a Big Bang event. This universe then undergoes two stages of expansion. In the first stage of expansion the universe expands until gravity slows down the rate of expansion. At this point the second stage of expansion begins. This second stage of expansion is fueled by a repulsive force. [96]

This theory fits with the recent discovery of the accelerated expansion of the universe and the newly discovered sponge like structure of the cosmos (see Figures 29 (key), 30,31, and 32).

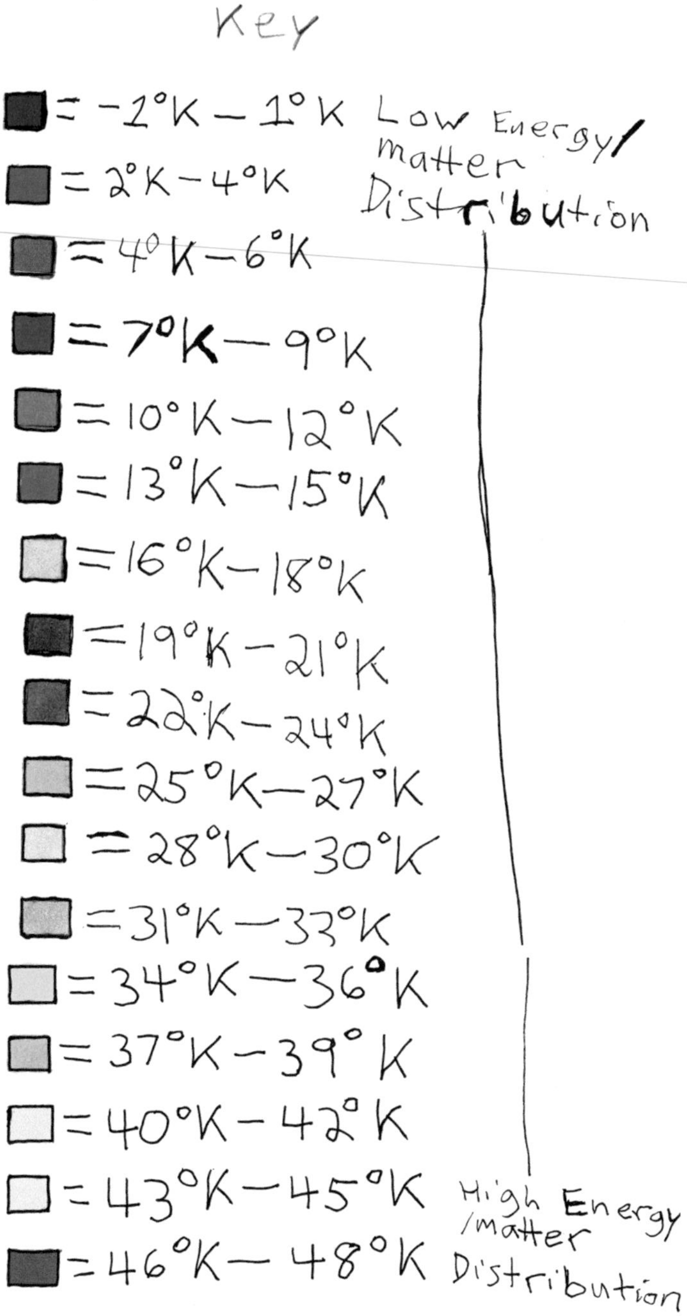

Key

■ = −2°K − 1°K Low Energy/ matter Distribution

■ = 2°K − 4°K

■ = 4°K − 6°K

■ = 7°K − 9°K

■ = 10°K − 12°K

■ = 13°K − 15°K

■ = 16°K − 18°K

■ = 19°K − 21°K

■ = 22°K − 24°K

■ = 25°K − 27°K

■ = 28°K − 30°K

■ = 31°K − 33°K

□ = 34°K − 36°K

■ = 37°K − 39°K

□ = 40°K − 42°K

□ = 43°K − 45°K High Energy /matter Distribution

■ = 46°K − 48°K

FIGURE 29

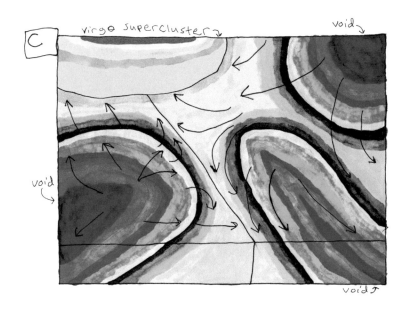

FIGURE 30

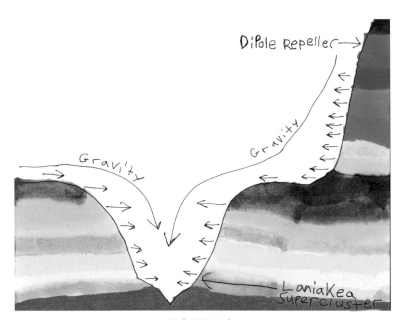

FIGURE 31

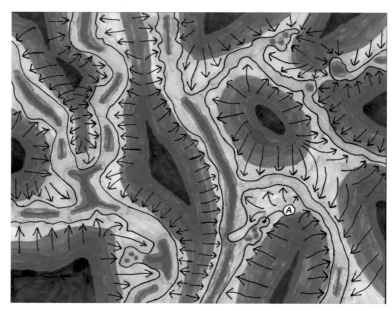

FIGURE 32

In Figure 32, point (A) marks where the Laniakea Super Cluster of galaxies is – our own Milky Way Galaxy is in this Super Cluster. Figure 33 shows a Cosmic Box depicting Cosmic Filaments. In Figures 34 (key), 35, 36, and 37; you see our universe as a four-dimensional spikey ball in cross section. Matt Parker tells us that: "Most people would describe circles and spheres as being smooth objects, but we have calculated that in higher dimensions they bristle with points." If our universe is a hypersphere as Carl Sagan and Alex Filippenko suggested, then our universe would have a spikey ball structure. [97.1 – 97.3]

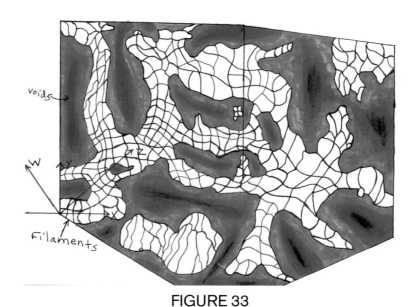

FIGURE 33

Key

$\blacksquare = 10 \times 10^{18} K - 10 \times 10^{15} K$

$\blacksquare = 10 \times 10^{14} - 10 \times 10^{12} K$

$\blacksquare = 10 \times 10^{11} - 10 \times 10^{9} K$

$\blacksquare = 10 \times 10^{8} - 10,000,000 K$

$\square = 10,000 K - 10 K$

$\square = 9 K - 2 K$

$\blacksquare = 1 K - ^{-}1 K$

$\blacksquare = ^{-}2 K - ^{-}10 K$

$\square = ^{-}11 K - ^{-}10,000 K$

$\blacksquare = ^{-}10,001 K - ^{-}10 \times 10^{6} K$

$\blacksquare = ^{-}10 \times 10^{7} K - ^{-}10 \times 10^{9} K$

$\blacksquare = ^{-}10 \times 10^{10} K - ^{-}10 \times 10^{12} K$

FIGURE 34

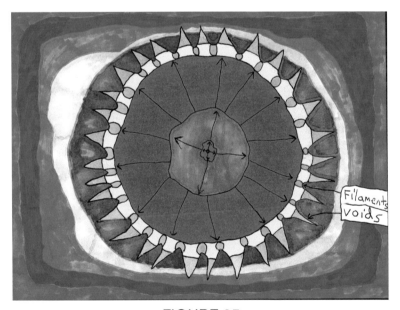

FIGURE 35

Some physicists and cosmologists have argued that our universe is closed and that there is **NOTHING** outside of our universe.

The scientists that believe that there is nothing outside of our universe make the following arguments. The first argument that they make is that they claim that the universe is infinite. And they define the "universe" as all that there is. [98]

First there is a problem with the definition of "universe". Cosmologists often talk about the **Observable Universe** and define it as the maximum part of the universe that we can observe. What this means is that we can only observe objects that are 14 billion light years away. We cannot see objects farther than this because this is the farthest distance light can travel in the time our universe has existed. This is our Cosmic Horizon. We cannot see beyond this boundary from Earth. We cannot just assume that there is nothing beyond the Cosmic Horizon. Assuming there is nothing beyond the Cosmic Horizon makes as much sense as looking to the horizon from the Empire State Building and assume that there is nothing beyond the horizon. Therefore, we cannot assume that we are seeing everything that is there. [99]

Matthew Francis says that: "For one thing, nobody sane believes that the universe we see encompasses everything: we would have to live in a very special place indeed if the universe is a sphere with us at the center, and based on observations that idea doesn't hold up at all." I discuss the Cosmic Horizon Problem in more detail in Section 2, Proof 1, Part 3. [100]

What this means is that when we are talking about the "universe" we are talking about either **THE WHOLE UNIVERSE** or **THE OBSERVABLE UNIVERSE**. **THE WHOLE UNIVERSE** is infinite and may indeed contain an infinite amount of finite **OBSERVABLE UNIVERSES**. **THE WHOLE UNIVERSE** containing an infinite amount of finite **OBSERVABLE UNIVERSES** is what cosmologists call **THE MULTIVERSE**. [101]

Now we can see why there are an equal number of physicists and cosmologists that say that there is something outside of our universe. In fact, Brian Greene, Carl Sagan, Stephen Hawking, Alan Guth, Clifford Pickover, and Lee Smolin are just a few scientists that say that there may well be something outside of our universe. [102.1 – 102.6]

One thought experiment that cosmologists often use to help them study the universe is called the Cosmic Box. Cosmologists plug all their observations of a portion of the night sky into a computer. The cosmologists put all the data from the portion of the sky that they studied into a box on the computer as a model for the whole universe.

One way that cosmologists use the Cosmic Box to study the large-scale universe can be seen in how they study how thermodynamics effects the development and evolution of the universe. [103]

Remember in the first section of this book we discovered that the outside of our universe had to be infinitely colder than our universe? Now let us briefly look at the thermodynamics of our whole universe in Figures 34 (key), 35, 36, 37 and 38.

A good example of this can be seen in Figure 36. In this picture we can see a small section of our Observable Universe. If we focus on just one Cosmic Void and the spike that is part of the void, we can understand the thermodynamic heat exchange that takes place between our Observable Universe and the environment in which it resides – The Whole Universe. The Cosmic Void Spikes act like chimneys. They allow heat from the galactic Filaments to rise into the fourth dimension into the larger universe while allowing the cold of the larger universe to cool the galactic Filaments. This is one way in which energy is transported between our Observable Universe and the Whole Universe. [104]

FIGURE 36

In Figure 37 we see that as the Cosmic Void Spikes grow larger their base widens. It is (in part) the negative pressure of the Cosmic Void Spikes that drives the expansion of our Observable Universe. In Figure 38 we can see where this fits into the charts of calculating Omega Mass and Omega Temperature of our universe. This chart helps us to understand how temperature, pressure, energy and mass are all related. Pressure can be seen as a form of energy and energy can affect how space warps (in other words it can affect gravity – the warping of space). [105]

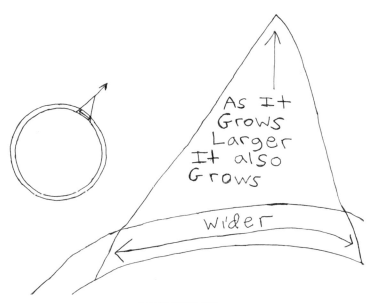

FIGURE 37

FIGURE 38

Remember in the First Section Figures 18 and 19 that showed the charting of Omega Matter and Omega Temperature? Figure 38 shows us Omega Temperature and Omega Pressure.

An interesting side note must be mentioned now. Remember in the First Section of this book where we came to the conclusion that thermal fluctuations (light) in a vacuum causes the folding of space (gravity). There are some scientists such as Joao Magueijo, John D. Barrow, and Lene Vestergaard Hau that believe that the speed of light is variable. [106.1 – 106.3]

Joao Magueijo explains that Einstein attempted to unify gravity with all other forces using a Kaluza–Klein metric. The Kaluza–Klein model of spacetime is five dimensional. This model has four spatial dimensions with time being the fifth dimension. In this model we live on a three-dimensional sheet occupying a four-dimensional space. Joao Magueijo says: "We are all flattened out so that we can never feel the larger space inside which we are embedded." [107]

Lene Vestergaard Hau has conducted experiments that shows the speed of light is indeed variable. Lene Vestergaard Hau used a Bose – Einstein Condensate to slow down light. She made a cigar shaped Bose – Einstein Condensate then fired a pulse of light through it. What she discovered when she did this was that the pulse of light not only slowed down but it also shrank significantly. The speed of light before it hit the Bose – Einstein Condensate was travelling 186,000 miles per second but after it entered the Bose – Einstein Condensate it slowed down to the speed of a bicycle. That is not all, the length of the light pulse went from being one to two miles long before it hit the Bose – Einstein Condensate to being compressed down to 0.001 microns in length. If light's speed were a constant, then it would stand to reason that a Bose – Einstein Condensate would not be able to affect it because light would have a constant speed no matter what medium it passed through. But that is not the case. See Figure 39. [108.1 – 108.2]

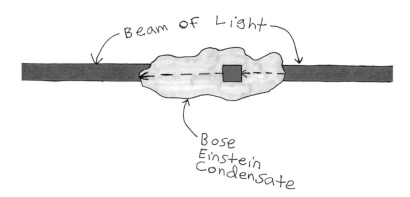

Beam of Light

Bose Einstein Condensate

FIGURE 39

If you cool down space, then the space folds in a way to slow down light. If you heat space up, then it might just fold in a way to make a pulse of light to move faster. Cosmic Voids have been seen to slow down light. [109]

The modern concept, idea, and definition of what a "universe" is has evolved through time. Once our most primitive ancestors thought that the small region of the planet that they inhabited was "all that there was". They did not believe that anything existed beyond the horizon. This was their definition of "universe".

Once tribes grew into kingdoms then empires a new definition of "universe" developed. The new definition of "universe" said that the small bit of continent that the empire and its neighbors inhabited was the "universe".

Then the Greeks discovered the curvature of the Earth and this brought about a revolutionary understanding that we inhabit a planet. Now the Earth became the "universe".

In 1915, when Einstein introduced his Theory of General Relativity to the world it was thought that our "universe" was the Milky Way Galaxy. [110]

Then Hubble discovered that there are other galaxies beyond our Milky Way Galaxy, and this changed our definition of "universe" yet again. [111]

PART 2

PROOFS OF A FOUR-DIMENSIONAL UNIVERSE DOWN TO ITS ATOMIC STRUCTURE

I contend that everything in our universe – down to its atomic structure – is four dimensional, because there is strong evidence for this, and we are now gaining the technology to test this hypothesis. Let us look at the evidence and how to test the hypothesis.

Proof 1: Part 1: The Expansion Rate of the Universe

So, if everything in our universe is four-dimensional, then why is it we only see three dimensions? Is there something wrong here? No, not at all. Once again, we will use classical physics to explain the discrepancy.

Our universe is expanding at 303,630,936 km/s and accelerating; therefore, it creates g-forces (ram pressure) on everything inside of it. You have 15 pounds per square inch of atmospheric pressure pressing down on your body, but you hardly notice it. This pressure is produced by the weight of the air pressing down on you. It is the velocity of our planet as it orbits around our sun and spins on its axis that drives the winds and the ocean currents. [112]

I calculated the g's that we are pulling as a result of the acceleration of our universe and I got either 30,982748.57 g's or 3,098,275.857 g's / $10m^4$. See Figures 41 and 42 for more details. [113.1 – 113.4]

Other people have also asked questions like, how many g's are we pulling due to the expansion rate of the universe and such questions are collectively known in cosmology as the Gravitational Paradox. [114]

Let's do another thought experiment. In the first episode of the TV series **Space 1999** called "Break Away", the Moon is blasted out of Earth's orbit by a series of huge thermonuclear explosions that take place at nuclear waste dumps that are situated on the Moon. For an hour, as the Moon is accelerated out of Earth's orbit, the people of Moonbase Alpha are subjected to enormous g-forces, which pin them to the floor of Moonbase Alpha. As the Moon decreases its acceleration rate, the g forces decreased as well. However, in our thought experiment, we will not decrease the Moon's acceleration rate nor increase it – we will keep the Moon's acceleration rate the same. [115]

So instead of being pinned for just an hour, the people of Moonbase Alpha will be pinned to the floor of Moonbase Alpha for one million years. Question: How will the people of Moonbase Alpha evolve? There are several changes that will occur.

First, the muscles the Moonbase Alphans need to move in the third dimension will atrophy and disappear, since the g–forces from the accelerating Moon will restrict their movements.

Second, since the Moonbase Alphans can no longer move in the third dimension, because of the g–forces imposed on them by their accelerating Moon, they have also lost the ability to see three–dimensional objects as well. What is the use of three–dimensional vision if you cannot move in three–dimensions? After all no threat can come at the Moonbase Alphans from the third dimension because it would be pinned by the g–forces like they are.

This creates a scenario where the Moonbase Alphans will eventually, (after a million years of evolution), perceive themselves to be only two–dimensional creatures because they have essentially become flatlanders. This is all a result of the g–forces caused by the acceleration of the Moon.

We are in a similar situation as the Moonbase Alphans in our though experiment. We cannot perceive four–dimensional objects, in the same way Moonbase Alphans cannot see three–dimensional objects. The g–force that is exerted by the acceleration of our universe pins us down in much the same way that the g force of the accelerating Moon pins the Moonbase Alphans. We are experiencing roughly 30,000 g's per 10 meters^4. This amount of g's is more than enough to prevent us from ever moving in or perceiving the fourth dimension (see Figure 40).

FIGURE 40

In addition, the people of Moonbase Alpha would be unaware of the larger universe they inhabit, because the g—forces they are under prevent them from ever looking up at the stars (see Figure 40). When the people of Moonbase Alpha circumnavigate the Moon, they will realize it is curved through a third dimension, but they can never measure that, because the g—forces they are under prevent them from moving up or down. Again, we are in a similar situation as the people of Moonbase Alpha. Both we and the people of Moonbase Alpha perceive our respective universes as "flat" because both we and they lack the ability to measure the curvatures in higher dimensions.

According to Newton's formula (F=ma), it does not matter how something accelerates, where it accelerates, why it accelerates, or when it accelerates all that matters when it comes to calculating the g—force exerted on the object being accelerated is the rate of the acceleration.

$$a_{Felt} = \frac{v^2}{r} + g$$

$$a_{Felt} = \frac{303,630,936 \ m/s^2}{10m} = 30,363,093.6 \ m/s^2 + 9.80 \ m/s^2$$

$$= 30,363,103.40 \ m/s^2$$

$$g's = \frac{a}{9.80 \ m/s^2}$$

$$\frac{30,363,103.40 \ m/s^2}{9.80 \ m/s^2} = 3,098,275.857 \ g's$$
$$per$$
$$10 \ m^t$$

FIGURE 41

$$g's = \frac{a}{9.80 \ m/s^2}$$

$$g's = \frac{303,630,936 \ m/s^2}{9.80 \ m/s^2}$$

$$30,982,748.57 \ g's$$

FIGURE 42

To effectively illustrate this fact let us do another thought experiment. If we launch a rocket from Earth its escape velocity would be 11 kilometers per second and if you were riding in the rocket you would be experiencing 3g's as you escaped the bounds of Earth's gravity. But if you were sitting on a balloon the size of Earth and this balloon was steadily inflated at 11 kilometers per second you would also be experiencing 3g's of force because the skin of the balloon on which you sit is traveling at 11 kilometers per second. (See Figure 43). [116.1, 116.2]

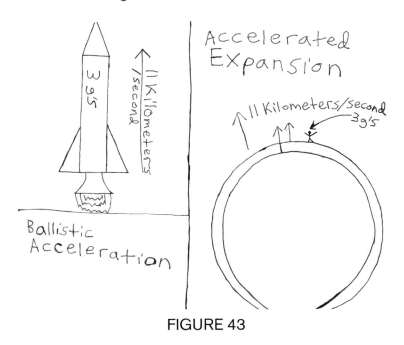

FIGURE 43

What would the Moonbase Alphans look like after one million years of evolving under these extreme g–forces? First, their skulls and bodies would be very misshapen by the g– forces – they would be flattened – like flounder (flat fish). The people of Moonbase Alpha would look horrifically deformed to us, and they would no longer be able to walk upright. Similarly, we would look horrifically deformed to a four–dimensional being and we would not be able to walk 'upright' in four–dimensions. (See Figures 44 and 45).

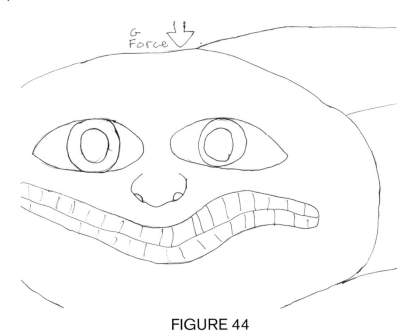

FIGURE 44

Moonbase
Alphan Face
(Note the Flattened Skull)

FIGURE 45

If our universe's acceleration rate is increasing, you would think that the g–forces would also increase, but this is not the case with our universe. Why is this not the case with our universe? This is because as the acceleration rate of our universe increases so does its mass – therefore, the g–forces remain the same. The structure that is increasing the accelerated expansion of our universe while it is increasing our universe's mass is a four–dimensional object called a Hyper Hole and we will be looking at these four-dimensional objects next.

In the following thought experiment, we can see why g-forces do not increase as our universe's acceleration rate increases. For example, if we have a van and increase its mass while at the same time increasing its acceleration rate, the resulting g–force we would feel while riding in the van would remain the same – it would be a constant (see Figures 46 and 48). However, if we increase the acceleration rate of the van but do not increase the mass of the van then the g–force we would experience while riding in the van would increase at the same rate as the acceleration rate (see Figure 47). But if we increase the mass of the van and leave the acceleration rate at a constant speed, then the g–force we would experience while riding the in the van would decrease, because the van would be increasingly more difficult to move with the same speed, until finally the van does not move at all (see Figure 49).

Edwin Abbott Abbott in the preface to the second and revised edition of **Flatland**, said that we may be living in a world with a fourth unrecognized dimension. [117]

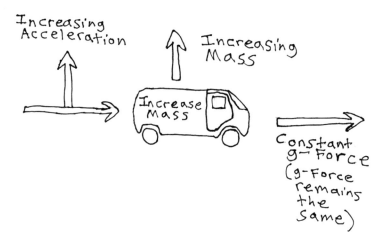

FIGURE 46

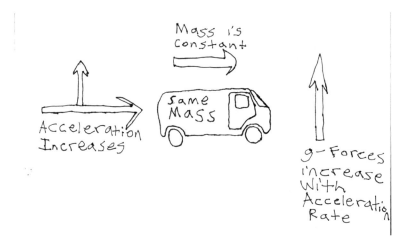

FIGURE 47

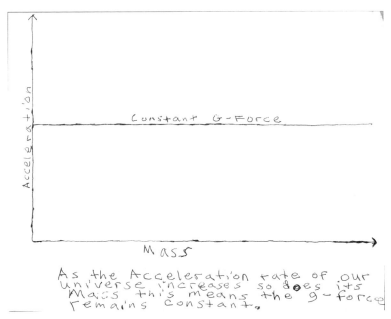

As the Acceleration rate of our universe increases so does its Mass this means the g-force remains constant.

FIGURE 48

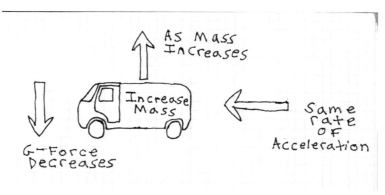

FIGURE 49

This can also be confirmed by observation. For example, if a given volume of space **NOW** on average contains five galaxies (in our 'local' space) and we look back (using Hubble Space Telescope) to the beginning of galaxy formation and compare it to **NOW** measuring how many galaxies a given volume of space has to see if the average galaxy density per volume of space that we see **NOW** has changed through time (getting larger or smaller) or does it remain constant. This paragraph may seem a little confusing to some people, so I will try to clarify it by using a little thought experiment in Section 2, Part 2, Proof 1, Part 3 The Cosmological Horizon Problem.

Proof 1: Part 2: Solving a Cosmological Crisis

The previous paragraph may seem a little confusing to some people, so I will try to clarify it by using a little thought experiment in Proof 1, Part 3.

Richard Panek describes a cosmological problem that is puzzling cosmologists like Adam Reiss. The problem has to do with the measuring of our universe's expansion rate. [118.1 – 118.4]

Astronomers and cosmologists have been looking for two numbers that could explain the development, evolution and history of our universe. The first number that they want to find is the deceleration parameter and the second number that they want to find is the Hubble Constant which tells us the current expansion rate of the universe.

Let us take a look at the deceleration parameter. As we have seen in Section 1 of this book our universe started to decelerate its expansion around 5 billion years ago before it suddenly started to increase its acceleration rate.

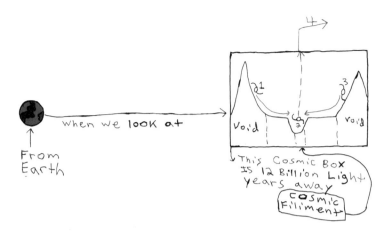

FIGURE 50

In Figure 50 we see astronomers on Earth looking at a Cosmic Box that is 12 billion light years away. The whole Cosmic Box is moving away from us at a H_0 (Hubble Constant) of 72 ± 5 kilometers per second per megaparsec (this is labeled 4 in Figure 50). There are three galaxies in the Cosmic Box that are labeled 1, 2, and 3. Galaxy 1 has a H_0 of 78 ± 2 kilometers per second per megaparsec because it is leaving a Cosmic Void Spike. As the galaxy leaves the four – dimensional Cosmic Void Spike it gains energy (momentum) that is why its H_0 is 78 ± 2 kilometers per second per megaparsec. The galaxy that is labeled 2 lies in a Cosmic Filament Valley that is moving away from us with the same H_0 as the Cosmic Box that it is in at 72 ± 5 kilometers per second per megaparsec. Galaxy 3 is moving from a Cosmic Void Spike into a Cosmic Filament Valley that has galaxy 2 in it. Galaxy 3 has a much lower H_0 than the other two galaxies because it is moving toward us and is therefore slightly blue shifted. Galaxy 3 has a H_0 of 69 ± 1 kilometers per second per megaparsec.

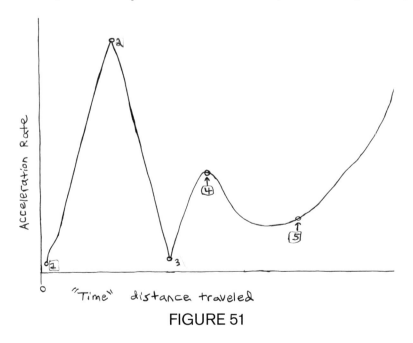

FIGURE 51

In Figure 51 we see how the universe's acceleration rate change with "time" (distance traveled). We will look at this in closer detail in Figure 52. In Figure 51 point 1 is when a stellar core in another universe forms a Hyper Hole (Hyper Holes will be discussed in the next section of this book). The stellar core starts out at relative rest in Figure 51. When this stellar core from another universe gets ejected from its parent universe into the Bose – Einstein Condensate Hydrogen, that surrounds its parent universe, it is traveling at a speed that is faster than light. It is the linear momentum of the stellar core's velocity that holds the stellar core together until it reaches point 2 in Figure 51. When the stellar core reaches point 2 it fully decelerates and as soon as it does it goes into a spontaneous explosive decompression (see the next section for more information). This spontaneous explosive decompression event is point 3 in Figure 51 and it is the Big Blast (Big Bang) event that started our universe. After the Big Blast (Big Bang) event our universe goes through an expansion phase that ends at point 4 in Figure 51. Point 4 in Figure 51 isn't only the end of our universe's first expansion phase it is also the beginning of a deceleration phase. This deceleration phase of our universe starts at point 4 in Figure 51 and ends at point 5 where a second phase of expansion begins. The H_0 (Hubble Constant) measurements that astronomers and cosmologists make will differ due to different factors.

FACTOR 1: Different stages in our universe's development will yield different H_0 (Hubble Constants).

FACTOR 2: Different large-scale structures of our universe will yield different H_0 (Hubble Constants) (see explanation of Figure 50).

An example of how different stages of our universe's development will yield different H_0 (Hubble Constants) let us now look at Figure 52.

In Figure 52 point 1 is the Big Blast (Big Bang) event that started our universe. Point 2 is the Cosmic Microwave Background Radiation (CMB) that is the afterglow of the beginning of our universe. The Cosmic Microwave Background Radiation (CMB) shows slight variations in temperature that eventually pushed matter into the filament and void structures that we see in the large-scale structure of our universe now. Point 3 of Figure 52 is where the first stars formed. These first stars are almost uniformly spread throughout the universe and the H_0 (Hubble Constant) is H_0 67 ± 5. These first stars are massive and they form neutron stars and Hyper Holes this is shown with point (a). Point (a) begins a line that shows the Hyper Hole production rate of the universe as it develops. Point 4 in Figure 52 shows where the first galaxies, Cosmic Voids, and Galactic Filaments form. The average H_0 (Hubble Constant) for these structures is H_0 68 ± 3. Point 4 is where we see the first signs of light doing thermodynamic work on the universe by impeding the expansion of the universe through the process of warping space into Cosmic Void Spikes and Cosmic Filament Valleys. At point 5 in Figure 52 is where the first galactic nuclei form thus producing the Quasars. The H_0 (Hubble Constant) for the Quasars is 69 ± 2. Point 6 in Figure 52 is where our universe's first phase of expansion reaches its peak and a phase of deceleration begins at this point. The H_0 (Hubble Constant) for this stage in the development of our universe is 72 ± 1. At point 7 in Figure 52 the deceleration increases as the matter in our universe collects in Cosmic Filament Valleys and in this way starts to form the modern large-scale sponge like structure of our universe and the H_0 (Hubble Constant) of these structures is 69 ± 2. Point 8 of Figure 52 is where the deceleration phase of our universe ends and the second phase of accelerated expansion begins and point (e) marks where the production of Hyper Holes has reached a point where these holes are pumping so much matter into our universe that it is the driving force in its accelerated expansion. The H_0 (Hubble Constant) for the galactic structures at this point is 67 ± 5. Point 10 in Figure 52 represents where we are now in our universe.

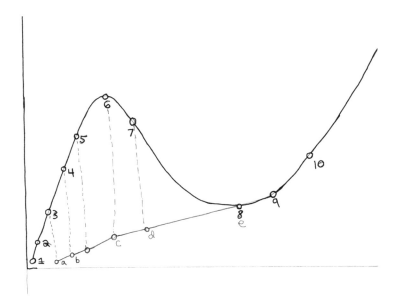

FIGURE 52

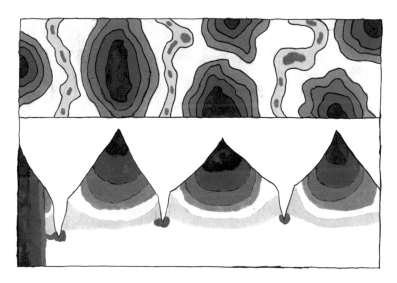

FIGURE 53

In Figure 53 we see two views of the Cosmic Web of Cosmic Void Spikes and Cosmic Filament Valleys. In the top view in Figure 53 we see a topological map like view looking from above the sponge like surface of our universe. In the bottom view we see a cross section of the top view.

Proof 1: Part 3: The Cosmological Horizon Problem

Let us explore the Cosmological Horizon Problem, mentioned earlier in this book, with a thought experiment. This thought experiment is similar to a problem that Yu L. Bolotin and I. V. Tanatorov faced that they call the (Cosmic) Horizon Riddle. This problem is more generally known as the Cosmic Horizon Problem. In Bolotin and Tanatorov's (Cosmic) Horizon Riddle they ask the following question: "Can two observers, in the same universe, with overlapping Cosmic Horizons pass information to each other regarding things outside of each other's Cosmic Horizon?" Neil deGrasse Tyson explains a little more about the Comic Horizon Problem. [119.1, 119.2]

In our thought experiment, however, there are three girls – Ethel from planet Earth, Juliet from planet Juno, and Chika from planet Achia (see Figures 62, 63, and 64). They all meet on Earth's moon (Luna) and participate in some activities like working at a Maid Cafe. See Figures 55, 56, and 57. Then they all travel back to their home planets using warp drive. But before each girl leaves the moon (Luna) they all promise to keep in touch with one another. But there is a problem. For example, when Juliet reaches her home planet of Juno and looks back towards Earth, she is seeing Earth as it WAS five billion years ago. Not as it is NOW (as Earth appears to us). Five billion years ago Earth's solar system was just forming, and Ethel was not even born yet. The light from Earth takes five billion years to reach the planet Juno. When Juliet traveled from Earth's moon (Luna) to her home planet, Juno, in a warp drive starship, she sped past the light just leaving Earth's solar system. The light from Earth's solar system that Juliet sped past will take five billion years to reach her home planet Juno. Actually, it will take longer than that because of the universe's expansion rate. The universe is in constant motion – it is not static. Therefore, if the girls only have access to radio communications then the promise that they made on Earth's moon (Luna) will remain unfulfilled because of the vast distances involved.

If Juliet sends a radio message to Ethel from Juno, then Ethel will receive the message in five billion years. From the perspective of Juliet, when Juliet looks towards Earth, she is looking at the Earth's past, but when Juliet sends a radio message to Earth, she is sending the message into Earth's future (see Figure 61).

In Figure 54 we see the Cosmic Horizon for each planet. Each planet can only see as far as 14 billion light years. In Figures 55, 56, and 57 the Cosmic Horizons are depicted as flat circles. This is very deceiving. In each figure all the circles represent a spherical Cosmic Horizon that each girl can observe from her home planet. All Cosmic Horizons are really, spherical patches on a four-dimensional hypersphere as depicted in Figures 58, 59 and 60.

We can see in Figure 56 a small circle around Earth. This represents Earth's 'now' which is very different than the planet Juno's 'now'. To Ethel on Earth, when she looks at Juno, she is seeing Juno as the planet Juno looked five billion years ago. Light takes five billion years to travel from Juno to Earth, therefore if some event happened on Juno and the people of Juno broadcasted the news to Earth, the broadcast would not reach Earth for five billion years. Please note that any messages sent from the planet Achia will never reach Earth because the planet Achia lies outside of Earth's Comic Horizon. The planet Juno is between planet Earth and planet Achia and that means that any event that occurs on Earth or Achia happens in Junos 'past'. From the point of view of a person on a planet looking up at the stars, they are not viewing the stars as they are 'now', they are viewing the stars as they were years, decades, hundreds of years, thousands of years, millions of years, or billions of years ago. The concept of 'now' is relative to the observer.

Bolotin and Tanatorov asked another question: "Can [Juliet] communicate to [Ethel] information that extends [Ethel's] knowledge of things that are beyond [her] Cosmic Horizon?" We shall see in Section 2, Part 2, Proof 1, Part 3 the answer to this question is: yes. [120]

In Figures 58 and 59 we see the circle that are depicted in Figures 55, 56, and 57 being translated to the curved four-dimensional surface of our universe in Figures 58, 59, and 60.

One question that you might be asking yourself by now is: "If the universe continues to expand will it ever 'pop'?" The answer is no, because more matter is being added to the skin of the universe as it stretches. We will discuss how more matter is being added to our universe in the next section.

In Figure 58 we see our universe in cross section. In this cross section we can see that as our spherical universe expands and grows larger the objects in our universe move progressively farther away from one another. This means that at one time in the past the three home galaxies of Ethel, Juliet, and Chika were much closer to one another (they even may have occupied the same galactic neighborhood) and over time all three galaxies will lie outside of each other's Cosmic Horizon and will forever drift farther and farther away from one another.

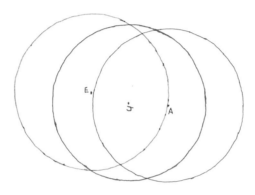

FIGURE 54

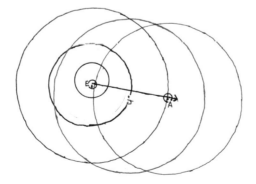

FIGURE 55

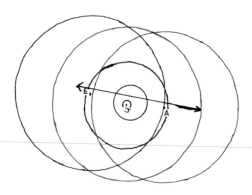

FIGURE 56

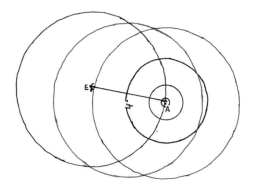

FIGURE 57

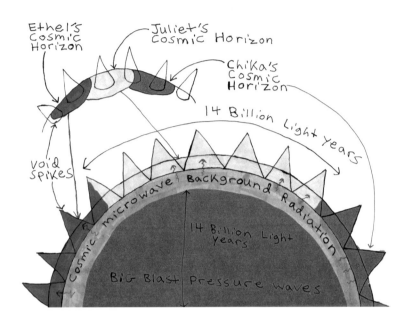

FIGURE 58

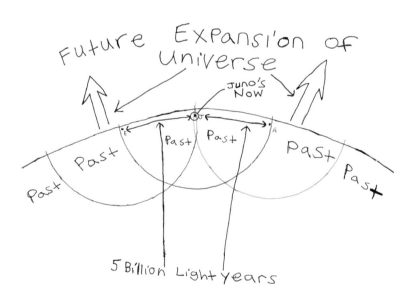

FIGURE 59

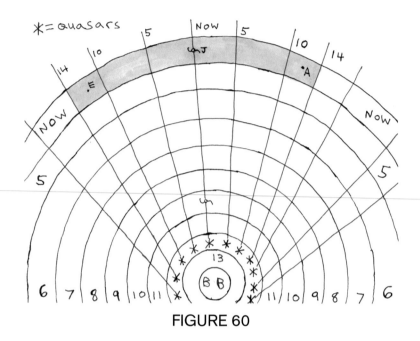

FIGURE 60

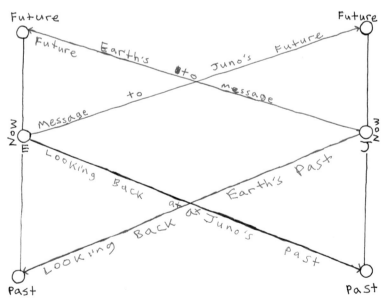

FIGURE 61

FIGURE 62

FIGURE 63

FIGURE 64

Another part of the thought experiment we started with Ethel, Juliet, and Chika is the fact that a "light year" is a very arbitrary form of measuring interstellar distances. For example, Chika's home world orbits a red dwarf star and this star is called Velcrome. Now in order for Chika's home world to be in Velcrome's "goldy locks" zone of habitability Chika's home world would have to be very close to the red dwarf star known as Velcrome. Chika's home world is 57.9 kilometers from Velcrome one orbit for Chika's home world is the equivalent of 88 Earth days. Earth takes 365 days to make one obit around the yellow dwarf star we call Sol. When the Earth makes one complete orbit around our sun (Sol) we call that a "year". Chika's "year" is much shorter than Ethel's "year" as a result therefore, Chika's measurement of a "light – year" is going to be much shorter than Ethel's measurement of a "light – year".

What about Juliet? How does she measure a "light – year"? Juliet's home world is a brown dwarf that orbits a yellow dwarf star like our sun. So, Juliet can measure a "year" in one of two ways. The first way that she could measure a "year" is to measure how long it takes her home world to make a single orbit around the brown dwarf. Juliet's home world takes 44 Earth days to orbit the brown dwarf. The second way that Juliet could measure a "year" would be to measure how long it takes the brown dwarf to orbit the yellow dwarf star. If the brown dwarf was 778.4 kilometers from the yellow dwarf star it orbits, then it would complete one orbit in 11.86 Earth years. Let us assume that Juliet uses this way to measure a "year", therefore, a "light – year" would be huge in comparison to Chika's and Ethel's measurements of a "light – year".

Out of all these arbitrary ways of determining a "year" and "light – year" which measurement should we use? The answer is none of them. One way of non – arbitrarily coming up with a measuring system that everyone can agree with is to look at the Cosmic Horizon Problem in the following way.

An Earth "light – year" is 9,460 billion kilometers. [121]

Which means that a light year for Juliet, for example, would be 112,195.6 billion kilometers (9,460 billion kilometers per year TIMES 11.86 years = 112,195.6 billion kilometers).

One way of making the measurement of a "light – year" less arbitrary is to look at how far we can see in the universe – our maximum Cosmic Horizon. We can see things that are 14 billion "light – years" away from us and that is our maximum Comic Horizon. How many kilometers is 14 billion "light – years"? From our vantage point in the universe we can look out and see things that are 132,440 billion kilometers away from us (14 billion "light – years" TIMES 9,460 billion kilometers per year = 132,440 billion kilometers).

So, what all of this means is that Ethel, Chika, and Juliet all can see out to a maximum distance of 132,440 billion kilometers. Now all you have to do to get a standardized "light – year" is to divide 132,440 billion kilometers into equal parts. For example, if we divide 132,440 billion kilometers by 20 then we get 6,622 billion kilometers and now we can use this as the new "light – year" that every civilization can agree with.

Proof 2: Hyper Holes

The term "Hyper Hole" was coined, by Clifford A. Pickover. [122]

In the first section of this book we showed that the outside of our universe is a vast, infinite sea of super cold Bose – Einstein Condensate Hydrogen. This Bose –Einstein Condensate Hydrogen, which is at a super low temperature, is also at a super low pressure. This means that as a super massive star goes through its evolution and reaches a stage where it has an iron core that it cannot fuse into more fuel, the star's core heats up. As it does so, it reaches a critical temperature and pressure, where upon it is ejected from our universe into the Bose – Einstein Condensate Hydrogen that surrounds our universe along a W axis (four dimensional) vector.

The mechanism for this can be seen in classrooms around the world. Get an empty glass bottle, a hard-boiled egg, some scrap paper, and a lighter or match. The egg represents the iron core of a super massive star and the inside of the bottle represents the outside of our universe. Place the egg on the mouth of the glass bottle. There are 15 pounds per square inch pressing down upon the egg, trying to push it into the bottle. But there also is another 15 pounds per square inch of pressure within the bottle trying to keep the egg out of the bottle. This is the normal state in our universe – there is a balancing of pressures that prevents our universe from flying apart at the seams. However, if that pressure is upset – for example, if there were more than 15 pounds per square inch of pressure applied to the egg, or if you put a burning match or piece of paper into the bottle in order to lower the pressure inside of the bottle – then the stellar core (egg) of a super massive star will pop outside of our universe (into the bottle) at very high speeds at or above the speed of light. If you cannot create a singularity by popping an egg into a bottle, then you cannot create a singularity by popping a stellar core out of our universe either (see Figures 65 and 66).

Let's look at a favorite experiment done by children to see why this is so. You have a milk bottle, an egg, and a match (see Figure 65). The egg represents a stellar core, the outside of the milk bottle represents our universe and the match inside of the milk bottle represents the low-pressure space that is outside of our universe. We all know what will happen when we place the egg at the mouth of the bottle it will just sit there because the outside pressure is not enough to crush the egg into the bottle – just as the pressure present anywhere in our universe is not strong enough (according to Einstein's 1939 paper) to crush anything into a singularity. But if we light the match and place it into the bottle then place the egg onto the mouth of the bottle the egg is both pushed (by the pressure on the outside of the bottle) and drawn into (by the low pressure created by the match) the bottle. [123]

A similar process works with stellar cores they are both pushed (by the internal pressure of the universe) and pulled (by the low-pressure space that lies outside of our universe) in such a way that (like the egg is never crushed) they do not create singularities of any kind. What they do create (in addition to a Hyper Hole) is a straight rifled Hyper Tube.

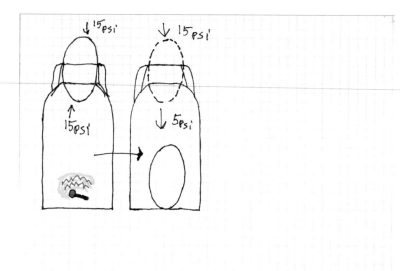

FIGURE 65

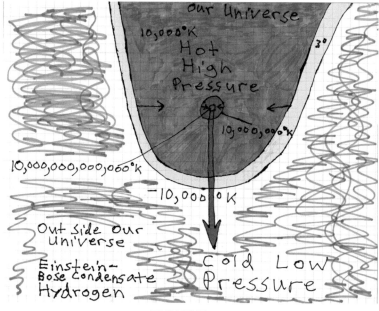

FIGURE 66

ROGER I. PARKER II

We know that everything in our universe is four–dimensional, so that means that when a stellar core leaves our universe, instead of creating a three–dimensional black hole it would create a four–dimensional hole called a Hyper Hole. This would also validate Einstein's claim, and the paper he wrote mathematically proving it, that black holes and singularities do not exist. [124]

Hyper Holes are created by Hyper Spheres (two spheres) when the Hyper Sphere (a stellar core) is torn from our universe. Each of the two spheres of the Hyper Sphere creates a hole (two spheres –two holes), that are separated by a fourth physical dimension (that is **NOT** time). This creates not one hole but two holes, one within the other separated by a fourth physical spatial dimension – a Hyper Hole. The outside hole is at low pressure, drawing matter to it. The inner hole is at high pressure, blasting matter out into the cosmos. There is a tube-like structure that leads to an identical hole elsewhere in another universe.

This arrangement can be understood in the following way. Imagine a crater that is located in the middle of a lake. Now if that were the case then the water in the lake would continually go into the pit. Now imagine that at the bottom of the same crater there is a firehose that is blasting water into the sky like a fountain. The large hole (the crater) is drawing matter (water) into it as the smaller hole (the firehose) that lies inside of the large hole (the crater) is blasting more matter (water) into the universe (lake). We can see this in Figure 67. [125]

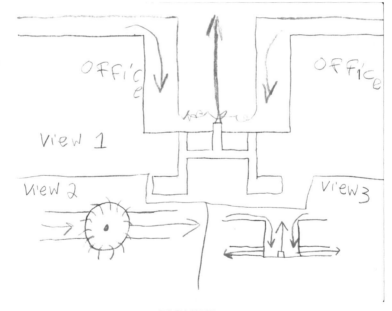

FIGURE 67

Niayesh Afshordi, Robert B. Mann, and Razieh Pourhasan also think that a four-dimensional stellar core is responsible for the origin of our universe. Afshordi, Mann, and Pourhasan say that our three-dimensional universe came from a four-dimensional imploding star (see Figure 68). In the scenario they put forth; is that a four-dimensional star in a four-dimensional universe (that has four spatial dimensions) implodes to form a singularity (a 'black hole'). [126]

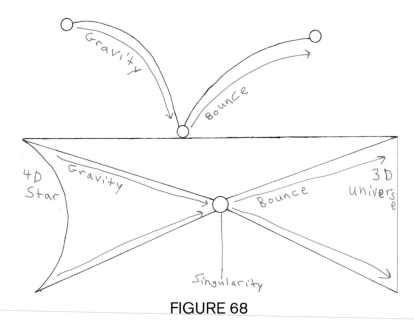

FIGURE 68

This singularity bounces much like a ball (see Figure 68). Gravity draws a ball to the Earth and gravity, likewise, compacts a star into an ever-smaller volume. A Tolman's oscillating universe is like a bouncing ball as it goes through an infinite number of cycles of Big Bangs that turn into Big Crunches. [127]

In Afshordi, Mann, and Pourhasan's scenario, however, there is only one 'bounce' when the four-dimensional star in the four-dimensional universe is crushed into a singularity before the singularity undergoes an explosive expansion (inflation) that forms our three-dimensional universe. [128]

Afshordi, Mann, and Pourhasan's claim that four dimensional 'black holes' exist runs into problems. Four dimensional 'black holes' and Hyper Holes are not the same thing. First a 'black hole' is a hole with a blockage – an obstruction in it that acts like a barrier preventing all passage forward. The singularity is the blockage in a 'black hole'. Another way of putting it a 'black hole' is a pit from which nothing can escape or move forward beyond singularity within it. However, wormholes and Hyper Holes have no obstructions impeding passage through them. They are tunnels that allow passage from one end to the other of the tunnel's tube.

Henry Norman states that a "singularity" is a place in a physics equation that is undefined or unbounded. Therefore, it does not necessarily refer to a real object. So that means that a "singularity" is just another mathematical abstraction that nobody understands. [129]

There is another problem with Afshordi, Mann, and Pourhasan's scenario and that is, if a universe that has four spatial dimensions produces a singularity that gives birth to our universe with only three spatial dimensions wouldn't that mean that when our universe produces a singularity it would produce a universe with only two spatial dimensions – a flatland universe? That does not sound very likely to me.

Hyper Holes form differently. In Figure 69 we see the stellar core that is ejected from our universe is hitting the Bose – Einstein Condensate Hydrogen. The iron stellar core was formed under high pressures, so it is like a spring under maximum scrunch. When the iron core is ejected from our universe it is traveling faster than light (traveling at warp speed). It is the ram pressure that this velocity exerts on the stellar core that keeps it together. Once this pressure is released by deceleration the stellar core experiences explosive decompression in the low-pressure environment of the Bose – Einstein Condensate Hydrogen.

When a stellar core leaves our universe, it creates a straight Hyper Tube because its spin is so fast (in the case of Neutron Stars) that the angular momentum it has when it leaves our universe forces it into a straight-line trajectory. The spinning stellar core creates a rifled Hyper Tube. Let us take a journey through the Hyper Hole – Hyper Tube system. Matter is drawn to the Hyper Hole's outer hole, where an accretion disc forms. The supper heated matter in the accretion disc is then drawn into the outer hole of the Hyper Hole, where it is accelerated to hyper light speeds (speeds greater than the speed of light) as it is spun by the rifled Hyper Tube. A heat exchange takes place in the Hyper Tube at this juncture, which draws in matter from the space outside our universe and injects it into the inner hole of the Hyper Hole to then be blasted into our universe at rifled relativistic speeds, like a fire hose (see Figure 71). It is this new material that is being blasted into our universe by Hyper Holes that is driving the accelerated expansion of our universe.

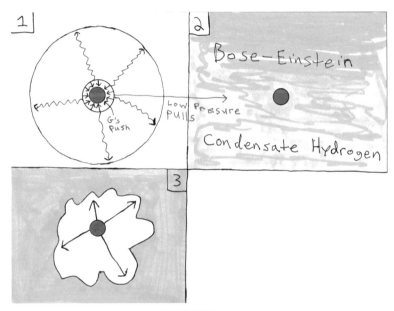

FIGURE 69

The cold inter–universal medium of Bose – Einstein Condensate Hydrogen is being blasted into our universe through these Hyper Holes. The Hyper Holes are pumping more matter into our universe, much like when you pump up a tire to inflating it. Our universe is inflating in the same way. In Figure 70, you see the steady deceleration of the universe due to cooling. This means that the thermal expansion of the universe is slacking off. The jerk – change in acceleration rate – and the increase in acceleration are both a result of Hyper Holes. An increase of Hyper Hole pumps pumping more matter into our universe is the reason for the ever-increasing rate of acceleration of our universe. The more Hyper Hole pumps there are the faster the rate that more matter is pumped into our universe.

Herman Bondi, Tommy Gold, and Fred Hoyle came up with a model of a steady state universe that expands. As the universe expands new matter is being created in some way, this is another way to express and interpret Einstein's Cosmological Constant. This type of Cosmological Constant gave the rate of new material being created and being added to the universe at a constant rate. [130]

Figures 71 to 76 show the structure of a Hyper Hole.

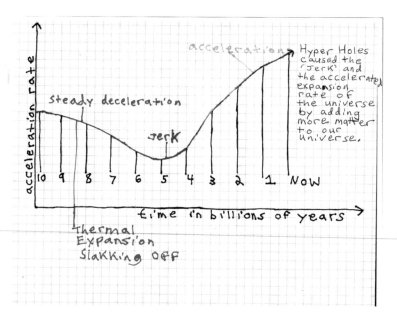

FIGURE 70

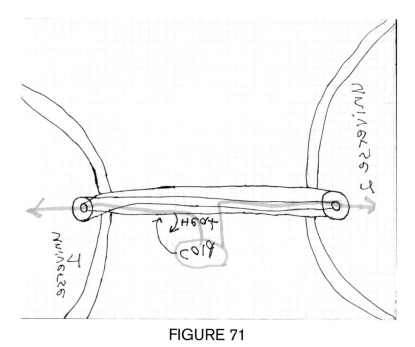

FIGURE 71

4 DIMENSIONAL OBJECTS
HYPER HOLES

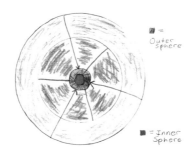

FIGURE 72

4 DIMENSIONAL OBJECTS
HYPER HOLES

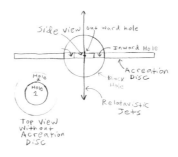

FIGURE 73

4 DIMENSIONAL OBJECTS
HYPER HOLES

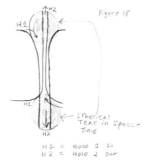

FIGURE 74

4 DIMENSIONAL OBJECTS
HYPER HOLES

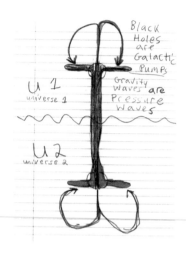

FIGURE 75

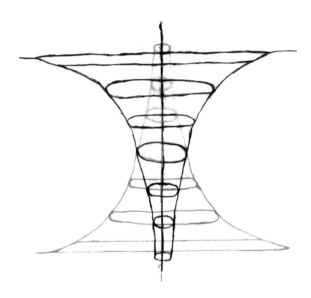

FIGURE 76

How do we know that the outer hole does not curve back on itself to form the inner hole? Simple observations of the relativistic jets that come from radio galaxies show us this is not the case. These jets blast out more material than that of the entire radio galaxy. For example, some radio galaxies measure only 100 thousand light years in diameter, but the jets they produce are more than 10 million light years long. Not only that, these jets can produce plumes that measure millions of light years in circumference. With more than 10 times the mass of such galaxies being blasted from these so called 'black holes', one wonders, where is all that matter coming from. Certainly, it is not coming from the galaxy itself. So where is it coming from, another universe perhaps? No, that would not be likely. Our universe is not unique, and therefore, if there is not enough matter in the galaxies within our universe to produce the jets then the same is true for any other universe.

Let us look at the structure of the Hyper Hole to see if this clears up the mystery of the missing matter. There can be no singularity present in any Hyper Hole. The outer hole of the Hyper Hole draws in matter and accelerates it to hyper light speed. [131]

Each hole is a Laval Nozzle that channels matter in one direction only into or out of our universe. [132]

Hyper Holes are similar to nested Klein Bottles as seen in Figures 77 – 82.

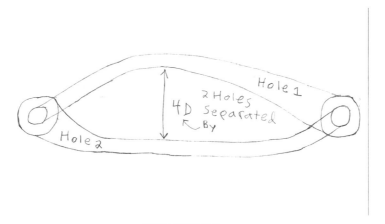

FIGURE 77

FIGURE 78

Clifford Stoll looked at a problem that Michael Spivak had in his book, ***A Comprehensive Introduction to Differential Geometry, Volume 1, Third Edition***. Clifford Stoll tried to solve the problem of a 'hole in a hole' that was introduced in Michael Spivak's book. What he found and proved is that the problem depicted in the book topologically was NOT a hole in a hole, it was instead two holes that were oriented at 90 degrees in respect to the central hole. See Figure 80. [133.1 – 133.4]

In his book **Full Color Illustrations of the Fourth Dimension, Volume 2** Chris McMullen describes four – dimensional stars and planets. [134]

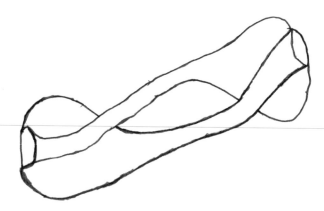

FIGURE 79

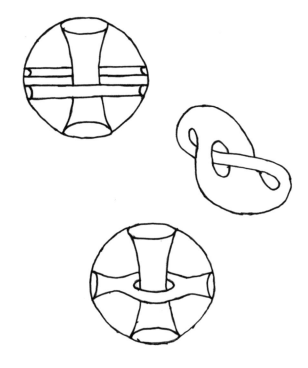

FIGURE 80

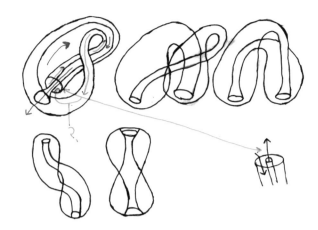

FIGURE 81

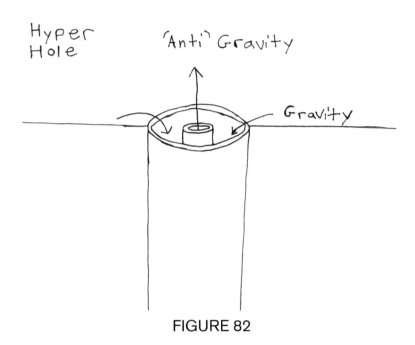

FIGURE 82

A stellar core when it leaves our universe creates a straight Hyper Tube because its spin is so fast (in the case of Neutron Stars) that the angular momentum when it leaves our universe is in a straight line. The spinning stellar core creates a rifled Hyper Tube. Let us take a journey through the Hyper Hole – Hyper Tube system. Matter is drawn to the Hyper Holes outer hole where an accretion disc forms. The supper heated matter in the accretion disc is then drawn into the outer hole of the Hyper Hole where it is accelerated to hyper light speeds (speeds greater than the speed of light) as it is spun by the rifled Hyper Tube. There is a heat exchange that takes place in the Hyper Tube at this juncture that draws in matter from the space that lies outside of our universe and injects it into the inner hole of the Hyper Hole to then be blasted into our universe at rifled relativistic speeds (this explains the missing mass that we mentioned earlier).

As we have stated earlier that nothing in our universe is static or stationary (see Section 1, Part 4) and John W. Macvey says that is why it makes more sense for there to be rotating 'black holes' than nonrotating 'black holes'. [135]

Some physicists do not think that Worm Holes (Einstein–Rosen Bridges) are possible because the singularity that would form a Worm Hole (Einstein–Rosen Bridge) would quickly close it (as seen in on the right in Figure 83. However, John W. Macvey says that the singularity that forms the Worm Hole (Einstein–Rosen Bridge) might not close it and instead allow safe passage around the singularity. Thus, forming a 'black hole' that drawls matter into it then forcing all that matter through the Worm Hole tube to the "white hole" exit where all the matter gets blasted out into another part of our universe or into another universe (as seen on the left in Figure 83). [136]

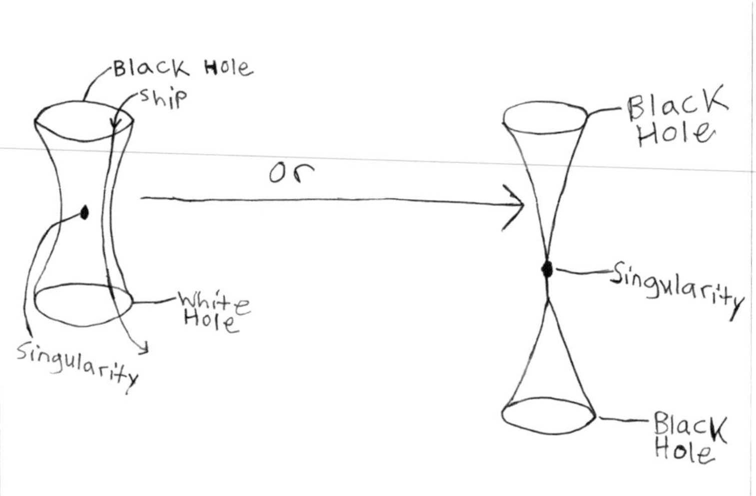

FIGURE 83

Hyper Holes act like pumps that draw in and expel matter, this makes a perfect mechanism for Einstein's Cosmological Constant. [137]

The very hot matter that is drawn into a Hyper Hole is cooled by the low pressure and temperatures that accelerate it to very high velocities. When this hot matter is accelerated and cooled by the very low pressure it draws in the matter that is outside of our universe into the inner hole of the Hyper Hole. The inner hole of the Hyper Hole the blasts this matter that comes from outside our universe into our universe. The other end of the Hyper Hole – Tube system lies in another universe. How do I know that? Simple, we have already discussed how the velocity of the expansion of our universe exerts pressure on everything in it (including stellar cores) and we also discussed how we can measure how much pressure that is. Also, we know that the temperature outside our universe must be colder than our universe and here too we discussed how to measure this. So, what is the pressure of the space outside our universe? We know both the temperature and the pressure of our universe. The colder something, like a gas, is then the lower its pressure is too. [138]

We have already explained that the temperature of the space outside of our universe is colder than that of our universe and we have also discussed a way of determining how cold it is and with this knowledge we can also determine its pressure because the lower the temperature of something, like a gas, the lower its pressure is – if we have an idea of the temperature we also can work out its pressure. [139]

But what happens to a stellar core once it leaves our universe?

Hyper Holes heat up our universe by drawing matter into their outer holes which results in spiral galactic structure. There is a cold region around the Hyper Hole, which is created when the Hyper Hole's inner hole blasts supper cool material into our universe at relativistic speeds ballistically, like a fire hose. When this material hits something, like a galaxy, it can ricochet off the object and thus produce plumes. [140]

The impact of such an event causes an enormous amount of heat and this heats up our universe too. The material that is injected into our universe eventually forms new galaxies. The supper hot regions created by the outer hole and the supper cold regions of the inner hole tilt the inner disc of our galaxy. The inner disc of our galaxy is located in the nucleus of the Milky Way and is tilted about 20 degrees in relation to the main galactic disk. [141]

This indicates that the Hyper Hole at our galaxy's center maybe acting like a warp drive engine. Miguel Alcubierre shows in a three-dimensional graph a warp bubble with a similar tilt relative to a spacecraft. [142]

As I have mentioned earlier, as a stellar core leaves our universe and enters the supper cold low-pressure regions of space that lie outside of our universe the low pressures do not have enough force to keep the stellar core together, so the unchecked internal pressure of the stellar core blasts the stellar core apart. Think of the stellar core as a spring under maximum scrunch. Once you have the spring squeezed as far as it will compress and then release the pressure it will spring back to its original shape. This is exactly what the stellar core tries to do.

There is another element to this scenario; the space that is outside of our universe must have something in it, right? What could that possibly be? Hydrogen that is cooled to Bose – Einstein Condensate temperatures would be a good logical guess. Our universe is composed mostly of hydrogen. Imagine a stellar core detonating in a sea of supper cold hydrogen. Hydrogen is very volatile and combusts very readily. So, what you would end up with is a very Big Blast and a lot of four–dimensional shock bubbles. Our universe is an expanding shock bubble. We see shock bubbles within our universe today. [143]

Alan Guth understands that the problem in constructing a scientific theory for the origin of our universe stems from a set of rules, called conservation principles. These principles trace their origin to the very roots of science itself and are its bedrock. [144]

Guth wonders were the universe comes from. Every culture has its own myths about the origins of the cosmos, but until recently, this question has been thought to be outside the scope of science. The Big Bang Theory says that the observable universe emerged from an explosion some 14 billion years ago, is widely accepted. Some versions of the Big Bang Theory suggest that all the matter in the universe was present from the start. Although the matter may have been in a different form, it was all there. The original Big Bang Theory describes the aftermath of the bang that created our universe. However, the theory makes no attempt to describe what "banged", how it "banged", or what caused it to "bang". Guth finally says: "Nothing can be created from nothing, we were always taught, so there was no hope for a scientific explanation for the actual origin of the matter in the universe". [145]

With our understanding of how thermodynamics shapes and molds, folds, and warps the fabric of our universe we can begin to understand how our universe formed.

Lee Smolin noticed a similarity between "black holes" and the beginning of the universe. He questioned whether a black hole was a singularity and stated: "Whether there is a real singularity is then a question that only a theory of quantum gravity can answer." [146]

Smolin went on to speculate that time does not end inside a black hole. If the singularity is avoided, then time can go on forever inside a black hole. If this is true, then that means that even though it is forever beyond the event horizon of the black hole, there is still something there, happening inside the black hole. The question now is, what is going on inside a black hole? [147]

Smolin notices a similarity between black holes and the Big Bang. He thinks that the formation of a black hole and the Big Bang describe a single event. For example, imagine a collapsing star forming a black hole. Within the black hole there is the star's core that is compressed to a very dense state. Our universe started in a similar dense state and then our universe started to expand from this dense state. Smolin thinks that the dense state of a stellar core inside a black hole and the dense state from which our universe started to expand from are one and the same. Smolin also believes that what is beyond the horizon of a black hole is the beginning of another universe. [148]

Smolin said that after a star explodes and it forms a black hole, its stellar core still exists inside the black hole's event horizon. We can't see beyond the event horizon of the black hole, but if we could see what is going on there then what would we see? Smolin thinks we would see a stellar core that is crushed to a very dense state explosively expanding to form a new universe. [149]

Smolin says that after the stellar core explodes inside the black hole, as it expands, this region of space may then develop much like our own universe did after the Big Bang. As this matter expands inside the black hole's event horizon, it may go through a period of inflation and become very large. If it develops suitable conditions, then stars and galaxies will start to form, and this new "universe" will become a copy of our own. Over time this new universe may produce life that gains intelligence, and while these beings look up at their night sky, they might be tempted to think that their universe was born in an infinitely dense singularity, before which there was no time. In reality, these intelligent beings would be living in newly formed region of space and time that was created by the explosion of a stellar core after a black hole formed around it in our part of the universe. [150]

Smolin goes on to say that the idea that a singularity could be avoided by an explosion, like the explosion from a stellar core after it passes the event horizon of a black hole, is very old. Smolin says that this idea goes back to the 1930's. At that time, cosmologists worried about the fate of our universe. According to cosmologists at that time, our universe started as a hot singularity that for some unknown reason expanded explosively. Then as our universe continued to expand and cool, it started to form stars, galaxies, and eventually us. The cosmologists of the 1930's speculated as to what might happen next, they theorized that our universe might slow down its expansion and then undergo a gravitational collapse into a singularity again. Many cosmologists today think that we live in what they call a 'Phoenix universe', which repeatedly expands and collapses, exploding again each time it comes close to becoming a singularity. This cosmic explosion, to expansion, to contraction and then, back to explosion cycle of a universe is called a **bounce**, because the repeated expansions and contractions of such a universe are analogous to a bouncing ball. [151]

Smolin concludes by saying that now we can apply this bounce hypothesis, not just to the universe, but also to every black hole within it. Smolin believes that we may be living in not just a single universe that is eternally passing through a recurring cycle of collapse and rebirth, but instead, we may be living in a continually growing community of universes, that are born from an exploding stellar core inside the event horizon of a black hole. [152]

John D. Barrow tells us that ever since the Big Bang Theory was first proposed scientists have asked: "what happened before the singularity that started Big Bang?" These scientists wondered if the Big Bang really did start with a singularity. [153]

In Figures 84 (key), 85, and 86 you see the Multiverse which is many Observable Universes tied together by Hyper Hole conduits. You will also see a stellar core being ejected into the Bose – Einstein Condensate Hydrogen, detonating and then start to form a new universe. In Figure 87 you see a stellar core detonating in the Bose – Einstein Condensate Hydrogen.

Remember in the previous section of this book where we talked about the Variable Speed of Light Theory. When our universe ejects an iron stellar core into the Bose – Einstein Condensate Hydrogen that surrounds it the iron stellar core will detonate as it explosively expands. The super cold Bose – Einstein Condensate Hydrogen will slow down any light produced by the detonating iron stellar core.

■ = ⁻10,000° – ⁻101° Kelvin
■ = ⁻100° – ⁻11° Kelvin
■ = ⁻10° – 1° Kelvin
□ = 2° – 3° Kelvin
■ = 4° – 9° Kelvin
■ = 10° – 100° Kelvin
□ = 101° – 1,000° Kelvin
□ = 1,000° – 5,000° Kelvin
■ = 5,001° – 10,000° Kelvin
■ = 10,001° – 100,000° Kelvin

FIGURE 84

FIGURE 85

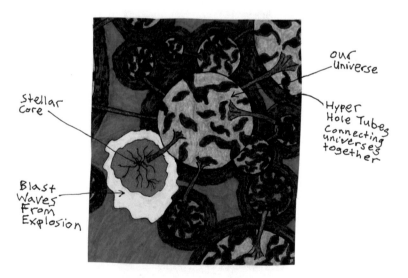

FIGURE 86

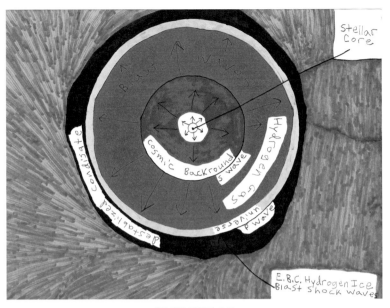

FIGURE 87

Many cosmologists say that magnetic fields emanating from the 'black holes' are responsible for relativistic jets. However, there are many problems with this theory. The first problem is that there is supposed to be a magnetic field associated with the 'black hole', but this has never been measured nor confirmed to exist. And even if there were a magnetic field, it would be too small and too weak to produce 10 million light year, long jets let alone create a column of twisting magnetic field lines. This can be seen in Figures 88 and 89.

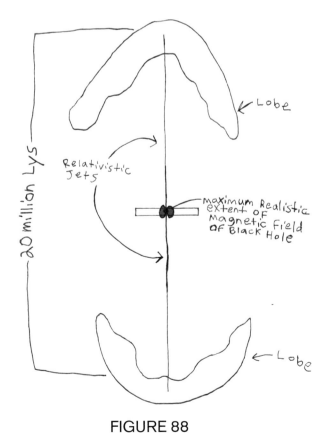

FIGURE 88

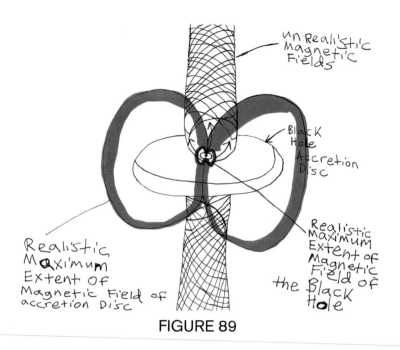

FIGURE 89

We can see this principle in our everyday lives. If we create a scale model of an active galaxy with its relativistic jet and if we represent the center of the galaxy with a small magnet, we will see that it would be impossible to make relativistic jets and the "bullets" of plasma that are associated with them (see Figures 90, 91 and 92).

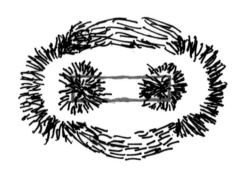

FIGURE 90

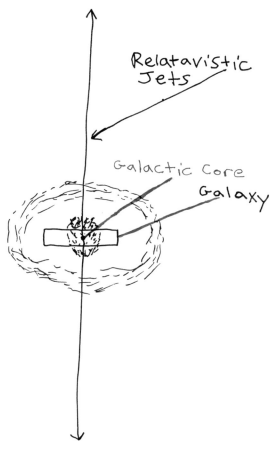

Relatavistic
Jets

Galactic Core

Galaxy

FIGURE 91

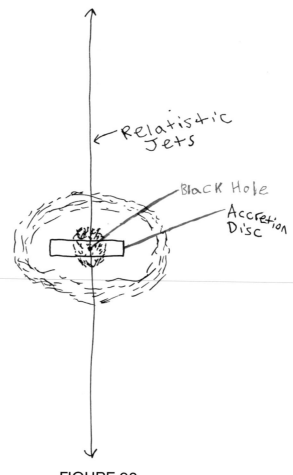

FIGURE 92

James Edward Biechier says that as galaxies grow, they enter four-dimensional space because the global three-dimensional curvature is not strong enough to pull the galaxy into our three-dimensional space. A halo forms around the galaxy. As the galaxy grows into the fourth dimension it pulls three-dimensional space up towards it. This is how the galactic halos are formed. When the galaxy pulls three – dimensional space up to it a gap is formed. When the galaxy is somehow pulled down into three – dimensional space halos are formed see figures 93 and 94. [154.1 – 154.2]

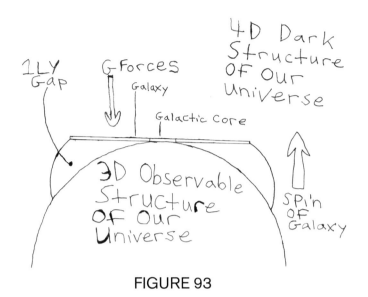

FIGURE 93

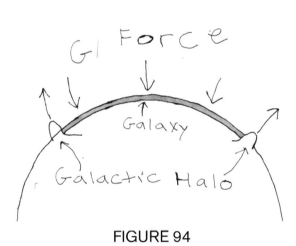

FIGURE 94

One last point to mention is Stephen Hawking claimed that 'black holes' can evaporate and that quantum fluctuations in a vacuum cause 'black holes' to evaporate. Hawking said that in quantum theory, fields can't be zero, even when in a vacuum. Hawking claimed that if the quantum fields were zero, then they would have an exact position at zero and an exact velocity at zero. Hawking believed that if the values for position and velocity were zero then this would violate the Heisenberg's Uncertainty Principle. According to the Heisenberg's Uncertainty Principle, the position and velocity of subatomic particles can't both be known for certain. Hawking believed that all fields must have vacuum fluctuations that keep us from knowing both the exact position and the exact velocity of a subatomic particle at the same time. The most popular way to think of vacuum fluctuations is to imagine two virtual particles that suddenly appear together at some point in spacetime, then move apart, and then come back together again annihilating each other in the process. When Hawking used the term 'virtual particle', what he meant is that although these particles could not be observed directly, their effects **can** be measured. He also claimed that these effects agree with theoretical predictions to a remarkable degree of accuracy. As an example, Hawking used the motions of a pendulum to illustrate his point. He said that a pendulums motion must be nonzero. [155]

First, any field can be zero by just not being there to begin with – according to basic logic. Second, the motion of a pendulum can never be nonzero because the motion of the Earth can never be nonzero. In other words, it is the motion of the Earth that causes some pendulums to be swinging eternally. As explained in Section 1, the Earth rotates eternally around its center of gravity. Jean Bernard Leon Foucault proved that a pendulum's motion is affected by the Earth's rotation and thus proved that our Earth does indeed rotate. The swing of a pendulum has nothing to do with any quantum fluctuations in a vacuum. [156]

As demonstrated with experimental evidence in the first section of this book, quantum fluctuations in a vacuum do not exist. What about the Heisenberg's Uncertainty Principle? Let us examine that next.

The Heisenberg Uncertainty Principle has been the excuse that physicists constantly use to try to magically conjure something from nothing. There is no experimental evidence showing the Heisenberg Uncertainty Principle to be true.

The Heisenberg Uncertainty Principle is only an artifact of our **TECHNOLOGY** and not a **NATURAL** phenomenon. In Section 2 Part 2 Proof 3 we will show that our universe is four-dimensional down to its atomic structure. So, that means that if we can only perceive only three dimensions adequately, that means we are only looking at 4% of what is truly out there in the universe.

Remember the Moonbase Alphans? They can't look at the stars above them because the g–forces they are under will not allow them to do so. So that means they are only looking at a very small fraction of what is around them on the Moon and are totally oblivious to what is going on around the Moon. But what if they developed the technology to see and move in the third dimension? If they could do this, then a whole new world would open for them. We are in a similar situation when it comes to the atomic structure of our universe. If we had the ability to see four–dimensional objects and map them properly, then the Heisenberg Uncertainty Principle would disappear instantly, because we could then see where everything was.

Hyper Holes are different from 'black holes' in the following three ways.

First, the radiation from Hyper Holes comes from friction (and other sources) and not from Hawking Radiation which supposedly comes from quantum fluctuations in a vacuum. The first experiment in this book proved that quantum fluctuations in a vacuum do not exist.

Second, Hyper Holes are stable and cannot evaporate.

Third, there is no singularity blockage in Hyper Holes.

This brings us to the question of, what is dark matter and dark energy? Dark energy and dark matter are the four–dimensional matter and energy that we cannot directly observe. Everything in our universe is four--dimensional down to its atomic structure, as we shall further explore in the next Part of this Section.

Proof: 3 Chemical Reactions in Four Spatial Dimensions

Theoretically speaking, it is possible that our universe's rate of acceleration is producing enough g's to prevent us from moving in the fourth dimension, let alone perceiving it. And that would mean Hyper Holes would theoretically exist. Gravitational lenses maybe further proof of our universe having four spatial dimensions. But is there anything we can do in the lab that will help us to experimentally prove that any of this is true? Yes, there is a way of experimentally proving all of this. We will use crystals to prove the existence of four spatial dimensions in our universe. It has long been known that some crystals have a four–dimensional spatial structure. [157]

For example, the four–dimensional structure of some crystals have been known since the late 1920's. [158]

This discovery added the four–dimensional crystals to the catalog of crystal families. One of the four–dimensional crystal families is very interesting and that one is the Hypercubic Family. [159]

Chris McMullen describes four – dimensional crystals and hypercubic crystal lattices by using hyper spheres to describe the atomic configuration of each hypercubic crystal lattice. [160]

Four–dimensional phase transitions have been reported in some four–dimensional crystals. If we could confirm that these four–dimensional chemical reactions really do occur, this would prove that our universe has a four–dimensional spatial structure.

A phase transition is when a substance, like water, changes from one state of matter to another. For example, when liquid water freezes and turns into a solid (ice), it is said to have undergone a phase transition. Scientists have discovered that phase transitions can occur in four-dimensional space and this would have to mean that our universe would have to be four dimensional (spatially) down to its atomic structure. [161]

We can see four – dimensional structures in the mitochondrial DNA of our cells. [162]

In Figures 95 (key) and 96 we see a two-dimensional designation of four-dimensional coordinates super imposed on a graph of a phase transition. See Figure 97 for a phase transition chart. For a definition of what Enthalpy is see Section 2, Part 4 of this book

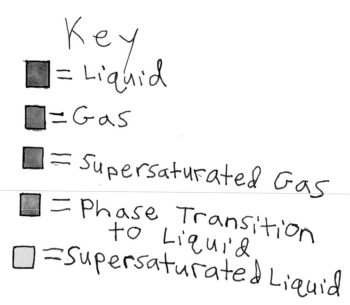

Key
■ = Liquid
■ = Gas
■ = Supersaturated Gas
■ = Phase Transition to Liquid
□ = Supersaturated Liquid

FIGURE 95

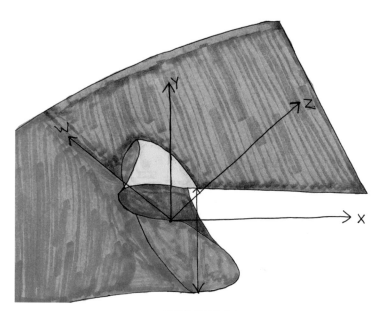

FIGURE 96

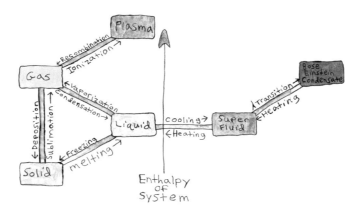

FIGURE 97

James Edward Biechier says that all sub atomic particles have a fixed four – dimensional spatial volume. This volume does not change. As a particle moves faster its volume contracts and gets pushed up into the fourth spatial dimension. When the particle moves slower than the speed of light the G – Force that is exerted by the expansion of the universe pushes the full volume of the particle into three – dimensional space. As we have discovered earlier in this book (see Section 1, Part 3). What we define as Energy is a Mass (in this case a sub – atomic particle) that is moving at a given Velocity. So, saying that any given sub – atomic particle has a constant volume is another way of stating the conservation of Energy and Matter. (See Figures 98 and 99). [163]

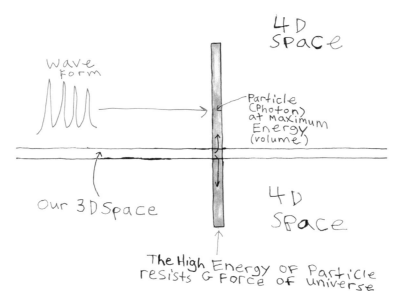

FIGURE 98

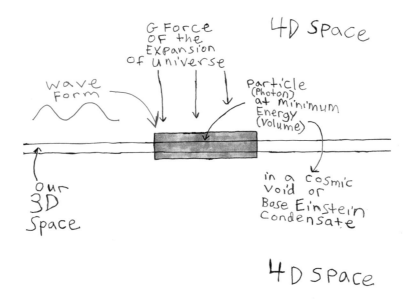

FIGURE 99

In 2008, it was reported that a urea–nonadecane crystal's structure has a four–dimensional spatial structure. [(164)]

As you can see, in any rendering of a tesseract, there are eight cubes and there are four axis – W, X, Y, and Z-axis (see Figures 100 to 108). Also, note that Cube 1 is the only cube we can observe or interact with. All the other cubes are unobservable, or **DARK**; therefore, we cannot directly observe them, nor can we directly interact with them either. This is where all the "dark" energy and matter lies. Now we know that all matter and energy in our universe is Four–Dimensional in nature and that "dark" energy and matter in our universe is the Four–Dimensional Energy and Matter that we cannot directly observe.

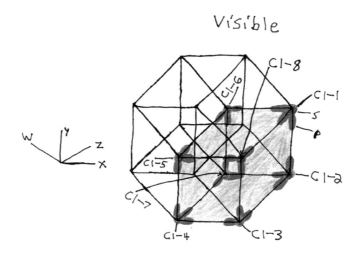

FIGURE 100

CUBE 1		COORDINATE LOCATIONS FOR CUBE 1
C1-1(XZY) =	C2-2(WZY) = C6-1(WXZ) = C8-1(WXY)	W (?) / X (33) / Y (15) / Z (?)
C1-2(YXZ) =	C2-3(YWZ) = C5-1(WXZ) = C8-2(YWX)	W (?) / X (33) / Y (7) / Z (?)
C1-3j(XYZ) =	C2-4(WYZ) = C3-2(XWY) = C5-2(XWZ)	W (?) / X (27) / Y (1) / Z (?)
C1-4(YZX) =	C3-3(WYX) = C4-4(WYZ) = C5-3(WZX)	W (?) / X (19) / Y (1) / Z (?)
C1-5(YXZ) =	C3-7(WYX) = C4-7(YZW) = C6-3(WZX)	W (?) / X (19) / Y (7) / Z (?)
C1-6(ZYX) =	C4-2(WZY) = C6-7(WZX) = C8-8(WYX)	W (?) / X (25) / Y (15) / Z (?)
C1-7(ZYX) =	C4-3(ZWY) = C5-7(WZX) = C8-3(WYX)	W (?) / X (25) / Y (7) / Z (?)
C1-8(XYZ) =	C2-8(WYZ) = C3-1(XWY) = C6-2(XWZ)	W (?) / X (27) / Y (9) / Z (?)

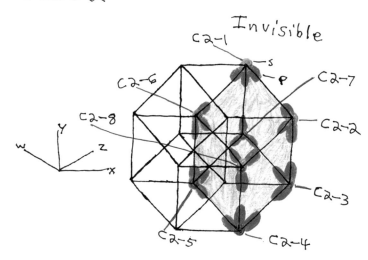

FIGURE 101

CUBE 2

COORDINATE LOCATIONS FOR CUBE 2

C2-1(ZYW) =   C6-6(XZW) = C7-1(XZY) = C8-6(XYW)   W (?) / X (27) / Y (21) Z (?)

C2-2(WZY) =   C1-1(XZY) = C6-1(WXZ) = C8-1(WXY)   W (?) X (33) / Y (15) / Z (?)

C2-3(YWZ) =   C1-2(YXZ) = C5-1(WXZ) = C8-2(XWY)   W (?) / X (33) / Y (7) / Z (?)

C2-4(WYZ) =   C1-3(XYZ) = C3-2(XWY) = C5-2(XWZ)   W (?) / X (27) / Y (1) / Z (?)

C2-5(XZW) =   C3-8(XYW) = C5-8(XZW) = C7-7(XYZ)   W (?) / X (21) / Y (7) / Z (?)

C2-6(YWZ) =   C3-6(XYW) = C6-8(XWZ) = C7-2(XYZ)   W (?) / X (21) / Y (15) / (?)

C2-7(ZYW) =   C5-6(XYW) = C7-2(XYZ) = C8-7(XYW)   W (?) / X (27) / Y (13) / Z (?)

C2-8(WYZ) =   C1-8(XYZ) = C3-1(XWY) =   W (?) / X (27) / Y (9) / Z (?)
                C6-2(XWZ)

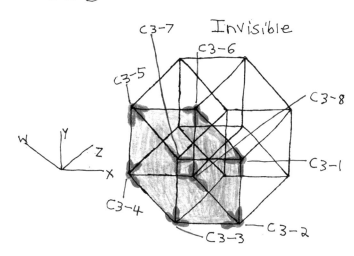

Cube 3

Invisible

C3-7
C3-6
C3-5
C3-8
C3-1
C3-4
C3-2
C3-3

W Y Z X

FIGURE 102

CUBE 3   COORDINATE LOCATIONS FOR CUBE 3

C3-1(XWY) =   C2-8(WZY) = C1-8(XYZ) = C6-2(XWZ)

C3-2(XWY) =   C2-4(WYZ) = C1-3(XYZ) = C5-2(XWZ)

C3-3(WYX) =   C1-4(YZX) = C4-4(WYZ) = C5-3(WZX)

C3-4(YZW) =   C4-5(YZW) = C5-4(ZXW) = C7-4(YZX)

C3-5(YWX) =   C4-6(YWZ) = C6-4(ZXW) = C7-5(YXZ)

C3-6(XYW) =   C2-6(YWZ) = C6-8(XWZ) = C7-7(XYZ)

C3-7(WYX) =   C1-5(YXZ) = C4-7 (WYZ) = C6-3(WZX)

C3-8(XYW) =   C2-5(YZW) = C5-8(XWZ) = C7-3(XYZ)

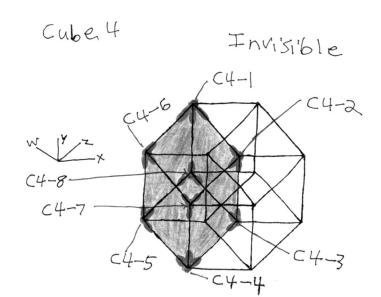

Cube 4

Invisible

C4-1

C4-6

C4-2

W Y Z

X

C4-8

C4-7

C4-3

C4-5

C4-4

FIGURE 103

CUBE 4

C4-1(ZYW) =

C4-2(WZY) =

C4-3(YWZ) =

C4-4(WYZ) =

C4-5(YZW) =

C4-6(ZWY) =

C4-7(WYZ) =

C4-8(ZYW) =

COORDINATE LOCATIONS FOR CUBE 4

C6-5(ZWX) = C7-6(ZYW) = C8-5(YWX)

C1-6(ZYX) = C6-7(WZX) = C8-8(WYX)

C1-7(ZYX) = C5-7(ZXW) = C8-3(WYX)

C3-3(WYX) = C1-4(YZX) = C5-3(WZX)

C3-4(YZW) = C5-4(WXZ) = C7-4(YZX)

C3-5(YWX) = C6-8(ZXW) = C7-5(YXZ)

C3-7(WYZ) = C1-5(YXZ) = C6-3(WZX)

C5-8(ZWX) = C7-8(ZYX) = C8-4(WXY)

Cube 5   Invisible

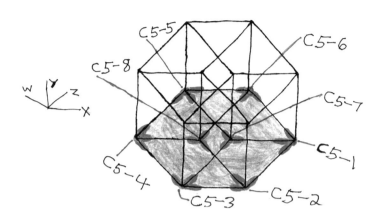

FIGURE 104

CUBE 5

C5-1(ZXW) =

C5-2(XWZ) =

C5-3((WYX) =

C5-4(ZXW) =

C5-5(ZWX) =

C5-6(XZW) =

C5-7(ZWX) =

C5-8(XZW) =

COORDINATE LOCATIONS FOR CUBE 5

C2-3(WZY) = C1-2(ZXY) = C8-2(XWY)

C3-2(XWY) = C3-2(XWY) = C2-4(WYZ)

C4-4(WYZ) = C3-3(WYX) = C1-4(YZX)

C4-5(YZW) = C3-4(YZW) = C7-4(YZX)

C4-8(ZYW) = C7-8(ZYX) = C8-4(YXW)

C2-7(ZYW) = C7-2(ZXY) = C8-7(XYW)

C4-3(ZWY) = C1-7(ZYX) = C8-3(WYX)

C3-8(XYW) = C2-5(YZW) = C7-3(XYZ)

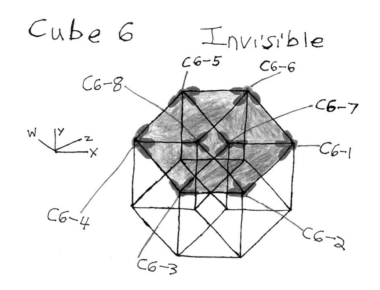

Cube 6   Invisible

FIGURE 105

CUBE 6

C6-1(ZXW) =

C6-2(XWZ) =

C6-3(WZX) =

C6-4((ZXW) =

C6-5(ZWX) =

C6-6(XZW) =

C6-7(WZX) =

C6-8(XWZ) =

COORDINATES FOR CUBE 6

C2-2(YZW) = C1-1(XZY) = C8-1(YXW)

C3-1(YXW) = C2-8(WYZ) = C1-8(XYZ)

C4-7(WYZ) = C3-7(WYX) = C1-5(YXZ)

C4-6(ZWY) = C3-5(YWX) = C7-5(YXZ)

C4-1(ZYW) = C7-6(ZYX) = C8-5(YWX)

C2-1(ZYW) = C7-1(XZY) = C8-6(XYW)

C4-2(WZY) = C1-6(ZYX) = C8-8(WYX)

C3-6(XYW) = C2-6(YWZ) = C7-7(XYZ)

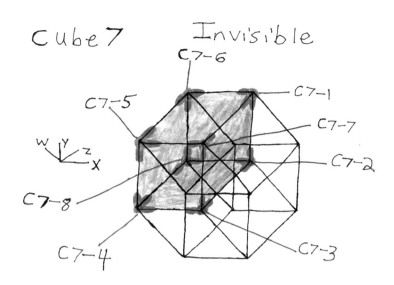

FIGURE 106

CUBE 7

C7-1(XZY) =

C7-2(ZXY) =

C7-3(XYZ) =

C7-4(YZX) =

C7-5(ZXY) =

C7-6(ZYX) =

C7-7(XYZ) =

C7-8(ZYW) =

COORDINATES FOR CUBE 7

C6-6(XZW) = C2-1(ZYW) = C8-6(XYW)

C5-6(XZW) = C2-7(ZYW) = C8-7(YXW)

C5-8(XWZ) = C3-8(XYW) = C2-5(YZW)

C5-4(ZXW) = C4-5(YZW) = C3-4(YZW)

C6-4(ZXW) = C4-6(YWZ) = C3-5(YWX)

C6-5(ZWX) = C4-1(ZYW) = C8-5(YWX)

C6-8(XWZ) = C3-6(XWZ) = C2-6(YWZ)

C5-5(ZWX) = C4-8(ZYW) = C8-4(YXW)

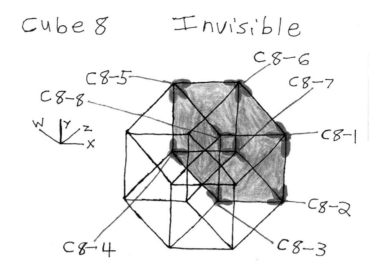

Cube 8   Invisible

FIGURE 107

CUBE 8	COORDINATES FOR CUBE 8
C8-1(YXW) =	C6-1(ZXW) = C2-2(YZW) = C1-1(XZY)
C8-2(XWY) =	C5-1(ZXW) = C2-3(ZWY) = C1-2(ZXY)
C8-3(WYX) =	C5-7(WZX) = C4-3(ZWY) = C1-7(WXY)
C8-4(YXW) =	C7-8(ZYX) = C5-5(ZWX) = C4-8(ZYW)
C8-5(YWX) =	C7-6(ZYX) = C6-5(WZX) = C4-1(WYZ)
C8-6(XYW) =	C7-1(XWY) = C6-6(XWZ) = C2-7(YZW)
C8-7(XYW) =	C7-2(YXZ) = C5-6(XZW) = C2-7(YZW)
C8-8(WYX) =	C6-7(WZX) = C4-2(WZY) = C1-6(ZYX)

You may be wondering why Cube 7 is listed as invisible. The reason is that it is on a W axis from Cube 1 (the cube that we are in and can observe). We cannot look in the w axis direction (where Cube 7 is located) and that means that we cannot observe it. However, in the following picture, you see that Cube 1 (where we are) and Cube 7 share a common portion that we **can** observe (see Figure 108).

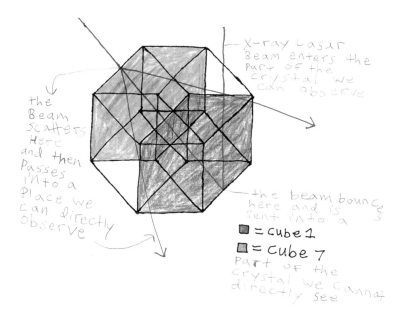

Handwritten annotations:

X-ray Lasar Beam enters the part of the crystal we can observe

the Beam scatters Here and then Passes into a place we can directly observe

the beam bounce here and is sent into a

■ = cube 1
▨ = cube 7

Part of the crystal we cannot directly see

FIGURE 108

We are only seeing one eighth of what is truly out there in our universe. There is little wonder why cosmologists say we can only see 4% of our universe. I recreated the diagram Alex Filippenko used in his lecture at Google to show this (see Figure 109).

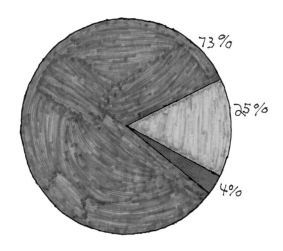

73%

25%

4%

FIGURE 109

A critical test to prove the four–dimensional atomic structure of crystals and the rest of our universe is to rotate a hypercubic crystal as we take x-rays of it. We do this by taking a fluoroscopy (x-ray movie) image of a hypercubic crystal as it rotates (see Figure 110). [165]

To test the theory that all matter in our universe is four–dimensional it helps to look at **ROTATING** cubes and Matt Parker tells us why. For example, Matt Parker describes a neat way we can examine the 2D shadow of a rotating 3D cube, it helps to color in two opposite squares faces of the 3D cube and leave the other four adjoining squares faces blank. That way you can track two of the squares as they rotate. Matt Parker coined the term 'hypoflatical' to describe an intelligent hypothetical being that is very flat. Instead of using his hypoflaticals, however, I will use the Moonbase Alphans. This rotating – shadow system can also be used to show a 3D cube to a Moonbase Alphan. As the 3D cube rotates next to the Moonbase Alphan, they can only perceive is its shadow, because all they perceive is a 2D universe. As the Moonbase Alphan watch the 2D shadow of the 3D cube, they would see one square get bigger as it draws closer to them and get smaller as it moves away from them. But if our Moonbase Alphans were to track two squares, one that is close to them (big square) and one that is far from them (little square), they will notice something that is strange to their perspective. As the 3D cube rotates, its 2D shadows also rotate. What the Moonbase Alphans see from their 2D perspective is that, the big and little squares keep passing through each other, as one square gets closer and the other square moves farther away from them. This would be outside the everyday experience of the Moonbase Alphans; for them no two squares could possibly pass through each other. What the Moonbase Alphans do not realize is that the squares are not going **THROUGH** each other, but rather in front and behind each other in a higher dimension. [166]

Parker tells us that if we project a rotating 4D cube's shadow into our 3D world, we would see something similar to when the Moonbase Alphans witnessed the shadow of a 3D cube. We know that as the 4D cube is rotating, it will cast a 3D shadow. This 3D shadow would look like a cube within a cube. This image is known as a tesseract. So, if we color in the two opposite faces of the 4D cube in and leave the other joining cubes transparent we will get an effect like the one described above. In the 3D shadow of a rotating tesseract (4D cube), we would see a large 3D cube which is the face of the tesseract closest to us. Inside this cube, we would see another smaller 3D cube, which is the face of the tesseract farthest from us. As the tesseract (4D cube) rotates, each 3D cube of its shadow looks bigger as it comes closer to us and smaller as it moves away. Again, as with the 3D cube's shadow to the Moonbase Alphans when they saw what appeared to be two squares passing through each other in the 2D shadow of a 3D cube: when we see the 3D shadow of a rotating 4D cube, the 3D shadow cubes will look to us as if they're going **THROUGH** each other as the 4D cube rotates, when in fact the 3D shadow cubes are going in front and behind each other in the fourth spatial dimension. However, this is something we cannot visualize. [167]

The Science Elf gives a superb video explanation on YouTube of Matt Parker's description of rotating 3D and 4D cubes. Verbal descriptions do not do justice to the concept of rotating 3D and 4D cubes – to fully grasp this concept, you must see it in action. This YouTube video demonstrates in a step by step manner how a rotating 3D cube casts a 2D shadow, and it shows the shadow–square faces of the 3D cube' shadow moving through each other as the 3D cube rotates. *The Science Elf* YouTube video also shows how a tesseract (4D cube) casts a 3D shadow and the video also, shows how the shadow cubes seem to move through each other as the tesseract rotates. In the video as the tesseract rotates, you can see how in the 3D shadow, the inner cube appears to become the outer cube, while at the same time the outer cube appears to become the inner cube. You will see the inner and outer cubes switch places again and again in the 3D shadow as the tesseract rotates. [168]

What this means is if we take an x-ray picture or an x-ray movie (Fluoroscopy) of a Hypercubic Crystal as it rotates, we will see the big and little cubes of the 4D Hypercubic Crystal moving through each other as they switch places, thus proving that our universe and everything in it is indeed four–dimensional (see Figure 110).

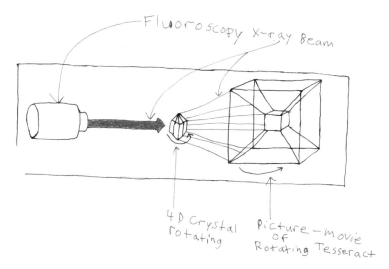

FIGURE 110

Similarly, Klein Bottles cannot exist in 3D space, but the Hyper Hole – that I previously discussed – is itself a Klein Bottle like object, which adds further proof that our universe is spatially four dimensional.

If it is even proven that our universe is indeed four – dimensional down to its atomic structure, then that would mean that zero- and one-dimensional objects are impossible. A singularity is a point like one dimensional object that is totally impossible to make in our real universe. Two dimensional objects are pseudo objects because even though we can see and interact with them we cannot touch them. Three dimensional objects are only partial objects because these are only parts of the four-dimensional universe that we inhabit (see figure 111).

Seven Dimensional objects		Hyper objects
Six Dymensional Objects		Hyper Objects
Five Dimensional objects		Hyper objects
Four Dimensional objects	Whole Objects	Hyper Cube
Three Dimensional Objects	Partial Objects	Cube
Two Dimensional objects	Pseudo Objects	Shadows
One Dimensional objects	Impossible	Singularity
Zero Dimensional Objects	Objects	

FIGURE 111

Einstein and Minkowski though that there were only three spatial dimensions, with time as another kind of dimension. But they never did define what "time" was. [169.1 – 169.3]

This viewpoint must be discarded, as it is no longer tenable.

Time must now be viewed as and perceived as a motion. **TIME** is really a motion of an object relative to other objects in the same space it occupies. How do we measure time? A day is how long it takes for the Earth to rotate once on its axis. A year is how long it takes the Earth to go around the Sun once. (Each planet has its own orbital period around a star – its own unique measurement of a year). This is an arbitrary way of measuring time. This way of measuring time is to measure the movement of the Earth relative to the Sun. Another way of measuring time that astronomers use is the Light Year. How long it takes light to travel in a year. This is less arbitrary, but it is still a measurement of an object's motion through space relative to other objects that occupy the same space. Atomic Clocks also rely on the measurement of an object's motion relative to other objects occupying the same space. [170]

TIME measures the movement of an object Relative to its environment. There is no way of measuring **TIME** without measuring the motion of an object Relative to its environment. Let us look at the history of measuring **TIME**. One of the first methods of measuring **TIME** was the Sun Dial which measured the apparent motion of the Sun in the sky. The Water Clock was the next way of measuring **TIME**. The Water Clock worked by measuring how much water leak out of a bucket with a hole in it. Candle Clocks measured the way a candle's wax melted as its wick burned. Mechanical Clocks measure the turning of gears and the motions of pendulums. Quartz Clocks measure how a quartz crystal oscillates when an electric pulse passes through it. Atomic Clocks measure how atoms oscillate when they absorb photons. [171]

Many works of Science Fiction use the concept of motionlessness to depict the stopping of 'time'. For example, in **Noein**, when time stops all motion stops and flying birds are frozen in mid–flight. [172]

Tom Jackson say that 'time' is basically a concept that nobody knows much about and nobody has adequately been able to define. [173]

Any event that occurs in our universe is localized by four spatial coordinates and one that tracks its motion through four – dimensional space.

The crystals prove that our universe has four spatial dimensions and not just three spatial dimensions.

And time must now be viewed as a quality (like color and texture) that all spatial dimensions possess. Things move in all spatial dimensions.

We have seen how the very expansion of our universe can prevent our perceiving and moving in the fourth dimension.

Einstein said that our universe has three spatial dimensions with x, y, and z spatial coordinates and one-time dimension with a single t–time coordinate. [174.1 – 174.3]

Einstein was wrong. Time never was the fourth dimension. The most advanced clock ever made – the Atomic Clock – measures the vibrations of a quartz crystal and synchronizes it with the vibrations of a cesium atom. A definition of one second is 9,192,631,770 oscillations (vibrations – or movements) of a cesium 133 atom. According to Khan Academy: "An Oscillator is an object or variable that can move back and forth, or increase and decrease, or go up and down, or left and right over and over." (175.1, 175.2)

According to Marshall Brian all Clocks keep track of 'time' by counting the 'ticks' (motions) of a Resonator. (176.1, 176.2)

A Resonator is a device or system that naturally oscillates at a given frequency. (177)

Therefore, the definition of "time" must be tied to the synchronized movements of two objects in space. "Time" is movement. It makes no sense to talk about the "beginning of time" any more than talking about the "beginning of motion". Time and motion are the same thing. Our sense of time is tied to our sense of motion.

Joanne Baker says: "In general relativity theory, the three dimensions of space and one of time are combined into a four – dimensional space – time grid, or metric." (178)

With all the evidence presented so far, we now know that our universe has at least four spatial dimensions. In fact, there is some evidence that our universe could have as many as six spatial dimensions. (179.1 – 179.3)

This is a major update to Einstein's theory of General Relativity. Now we can correct his equations by, for example, replacing "time" with distance traveled. "Time" and distance traveled are interchangeable in Newtonian Physics now they are interchangeable in Einstein's Theories of Relativity as well. (180)

In this way "time" is transformed from a Scaling Quantity to a Vector. A Scaling Quantity has magnitude, but it has no direction. A Vector like motion (and time can only be defined in terms of some motion) has a magnitude, speed (velocity), **and** direction. This must be changed to a grid of four – dimensional space only. (181)

Also, the incorrect view that dimensions that are not spatial must now be discussed. For example, in J. Divahar's representation of "four-dimensions" we can clearly see the problem of calling "time" any dimension. In his representation of "four-dimensions" he depicts a three-dimensional cube. As we have seen before, a cube only has three axis (x,y,z). J. Divahar then places a paper in the middle of the cube and calls it "four-dimensional" – he is absolutely wrong! Because no matter how many times you wiggle your 3D paper in the three-dimensional box you are still dealing with **three-dimensional axis not four.** It is like me saying that when I wiggle my left hand that is one dimension and that my right-hand wiggling is a totally separate dimension. What utter nonsense! (182)

In Figure 112 we see a representation of J. Divahar's "four-dimensions" and we see clearly that it is not only three-dimensional we also see that no matter how many times you wiggle the plane that is inside of the cube it will never be anything other than three-dimensions.

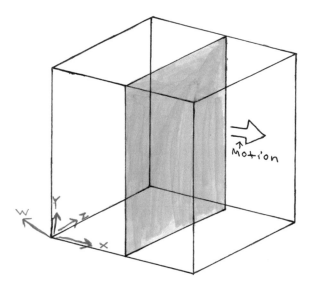

FIGURE 112

Einstein's Relativity needs to be revised in the following ways.

First, space (NOT TIME – there is no "spacetime continuum") is not flat it is instead curved four – dimensionally by thermodynamic forces.

Second, gravity is not a force, but instead it is a macroscopic effect of the electromagnetic – photovoltaic weak force. This has been experimentally proven in Section 1 of this book. The electromagnetic – photovoltaic weak force is most responsible for the warping of space (NOT TIME) that we call gravity.

Third, gravitational effects can slow down motion ("time" can only be defined as a motion).

Matt Parker tells us that the higher–dimensional cubes (the 7D, 6D, 5D, and 4D cubes) also project down into lower dimensions, but only their shadows project down. According to Parker, trying to picture anything past 5D objects requires so many interpretations of shadows–of–shadows that the resulting image is just not that useful. This means is that if we had a transparent 7D cube, for example, that we want to take a picture of. When we take a picture of it. The light will pass through the 7D cube to cast a 6D shadow, which then casts a 5D shadow that then casts a 4D shadow that then casts a 3D shadow that then casts a 2D shadow which becomes a quite messy picture (see Figure 113). [183]

For example, Brian Greene describes a six – dimensional object and shows a picture of it in his book **The Elegant Universe**. Brian Greene calls this six – dimensional object's shape a Calabi–Yau space. This shape is a complete mess. Brian Greene then describes ordinary three-dimensional doughnut like objects as Calabi–Yau objects. Is the six–dimensional object that he shows supposed to be a doughnut like object? If so then it is not rendered very accurately. [184]

Eugene Khutoryansky describes a way to draw 4, 5, 6, and 7 dimensional objects accurately. This is done by a "doubling the points" method. In her video she accurately rendered 3, 4, 5, 6, and 7 dimensional cubes. [185]

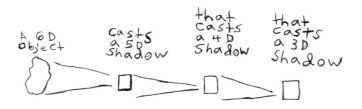

FIGURE 113

One last issue must now be addressed – that of so called "Time Crystals". Some quantum theoretical physicists, such as Frank Wilczek, talk about "Time Crystals" without any understanding of Geology, crystal formation and crystallography. If such scientists did have a complete understanding of these things, they would not make such a grievous error in thinking that you can hold time in your hand. Frank Wilczek and other scientists wrongly believe that matter at its resting state is motionless. We have seen in the first section of this book that this notion is incorrect. Everything in our universe down to its atomic structure is in motion. Even the atoms that make up a brick wall are in motion jiggling back and forth. Frank Wilczek claims to have made a "new type of matter" that moves at its ground state. We have seen that the ground state of our planet Earth is in motion around the sun (Sol). So that means that Frank Wilczek's claim of "time crystals" is a fantasy. I would like him to use one of these "time crystals" to go to the Jurassic Period and bring back a baby sauropod and have medical confirmation that it is indeed a dinosaur then send the baby dinosaur back to the Jurassic period. Extraordinary claims require extraordinary proof. [186.1 – 186.3]

Proof: 4 "Faster-Than-Light" Communications

Let us look at how such faster-than-light communications might be able to take place.

In the Figure 114, we see two planets of equal mass co-orbiting one another. Let us say they are about the same distance from one another as the Earth and the Moon. Light takes about two minutes to reach the moon and it would take about the same length of time for a beam of light sent from one planet to reach the other planet.

The reason that light takes two minutes to get from one planet to the other in our example, is that the light is traveling only three–dimensionally from one planet to the other and in doing so, it must climb out of one gravity well and go down the side of the other gravity well. However, if the beam of light were to travel four–dimensionally, then it would be able to avoid both gravity wells altogether and thus get from one planet to the other in a fraction of the time it would take if it had traveled three–dimensionally.

When light travels four–dimensionally between the two planets, it arrives at its destination almost instantaneously. Light traveling three–dimensionally is traveling at the same speed as the beam of light that is traveling four–dimensionally. The reason that the four–dimensional beam of light gets to its destination sooner is because it has a lot less terrain to cover.

For example, a good analogy for this would be to think of light traveling three—dimensionally is like a car and light traveling four—dimensionally is like an airplane. Even if the car and airplane travel at the same speed to the same destination, the airplane will always arrive before the car. Why? Because, the car must drive around mountains and up and down hills and valleys, find ways across rivers and avoid other obstacles. The path that the car takes avoiding obstacles to its destination is a very curvy, winding path and not a straight line. The shortest path between two objects is a straight line. Only the airplane can travel in a straight-line arc to its destination.

Stephen Hawking and Leonard Mlodinow stated that Einstein's General Theory of Relativity was a new theory of gravity that was nothing like Newton's theory of gravity. Unlike Newton, Einstein proposed that spacetime was not flat, as had been assumed previously, but instead it was curved by the mass and energy within it. [187]

Stephen Hawking and Leonard Mlodinow went on to say that picturing the curvature of the Earth's surface is a good way to think about this curvature. Hawking and Mlodinow compared the Earth's curvature to that of a four—dimensional sphere. The Earth's surface is a curved two—dimensional space. A hypersphere's (a 4D sphere's) surface is a curved three—dimensional space. The geometry of curved surfaces such as the earth's surface or the surface of a hypersphere is not the Euclidean geometry, we are familiar with. For example, in Euclidean geometry the shortest distance between two points is a line; however, in curved spaces or curved surfaces the shortest distance between two points is along a curved path called the great circle. On the Earth, the great circle that is along the Earth's surface and its center coincides with the center of the Earth. An example of a great circle on the Earth is the equator. [188]

Stephen Hawking and Leonard Mlodinow explained that we travel part of a great circle every time we fly from one city to another. For example, let's say that you wanted to travel from New York to Madrid. These two cities are almost at the same latitude. If the Earth were a flat surface, according to Euclidean geometry, the shortest route would be to head east in a straight line. If the Earth were a flat surface and you flew from New York to Madrid, then you would arrive in Madrid after traveling 3,707 miles. However, the Earth is not a flat surface – it is curved into a third spatial dimension (non—Euclidean geometry). Therefore, due to the curvature of the Earth, there is a curved flight path that you can draw on a flat map that looks like a longer distance, but this curved flight path is really the shortest route possible. If you used the curved flight path to fly from New York to Madrid, then you would arrive in Madrid after traveling only 3,605 miles if you follow the great—circle route. The great circle route from New York to Madrid is to first head northeast, then gradually turn east, and then southeast. The difference in distance between a straight-line route drawn on the map and a curved route following the great circle is due to the earth's curvature, and this is non—Euclidean geometry at work. The differences between Euclidean geometry and non—Euclidean geometry are well known to airlines; therefore, airlines arrange for their pilots to follow great—circle routes whenever practical. [189]

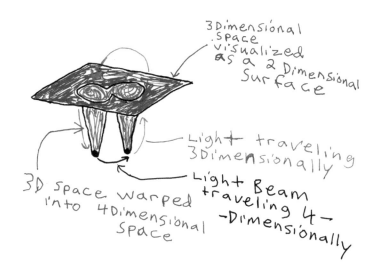

FIGURE 114

Stan Gibilisco tells us there may be a way to quickly send messages across vast interstellar and intergalactic distances. The way he thinks this could be done is to transmit radio waves into the fourth spatial dimension and that way the radio waves would appear to travel faster than the speed of light. But, in reality they would be traveling at the same speed as light, however, they would arrive faster because they are moving in a four–dimensional curved path. [190]

In September 2011 a team of researchers announced that they had detected faster than light particles at a lab in Gran Sasso, Italy. Another group of researchers found more evidence for faster than light particles at CERN. The scientific community investigated these claims. They concluded in June 2012 that all the claims of faster than light particles can be attributed to faulty measuring instruments. [191]

However, what if their instruments were not faulty. If that is the case, then we may be able to create a real time communication system for space craft. Right now, if we launch a group of astronauts to Jupiter it would take about 43 minutes to send a message from Earth to Jupiter. This type of delayed communications is not all that bad when we consider the communications delay would be 4 years if we launched any ship to Alpha Centauri. In the movie **Star Trek IV: The Voyage Home** the bridge crew of the **Enterprise** travel from Vulcan to Earth in a captured Klingon vessel. [192]

In the **Star Trek** universe, the planet Vulcan is 16 light years away from Earth. [193]

As the bridge crew of the **Enterprise** begin traveling to Earth, they receive an emergency message from Earth. If the message was sent by radio, then the message would never reach the **Enterprise** crew when they leave Vulcan. However, in the hour it takes the **Enterprise** crew to travel to Earth at warp speed they do receive a message. The only possible explanation for this is that the message was sent in a four-dimensional arch from Earth to Vulcan. If that is the case, then it is theoretically possible to have almost real time communications between Vulcan and Earth (see Figure 115).

This makes sense because warp drive is theoretically possible and what good is warp drive without four dimensional (faster than light) communications to go along with it. [194]

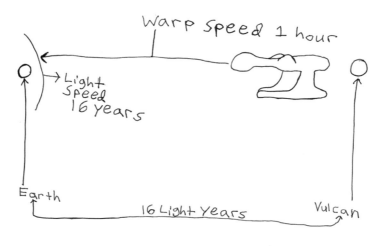

FIGURE 115

Proof: 5 Gravitational Lenses

In the first section of this book, you learned that gravity is not a force. Now I am going to clarify that. John Gribbin explains that what we think of as gravity is really the bending and warping of four–dimensional spacetime (three of space and one of time). Spacetime is bent or warped by the presence of material objects. Gribbin says that this warping of spacetime that we call "gravity" can bend the path of a beam of light. [195]

John Gribbin goes on to explain that, Einstein said in effect, that gravity isn't a "force" because in four–dimensional space time, objects are following a path of least resistance the equivalent of a straight line, through the curved portion of spacetime. The effect is the same for all objects in spacetime, even light is affected by the curvature of spacetime. All objects have mass, and objects with larger masses bend spacetime more than objects with less mass. When an object distorts and bends spacetime it is known as a gravitational field of force. [196]

Gribbin tells us that the new mathematics that Riemann developed became a vital tool for Einstein's work. Einstein used Riemannian geometry to investigate a four – dimensional model of the universe that is an analog of a sphere, a 'hypersphere'. The surface of our hyperspherical universe is the three dimensions of space that we experience directly. These three–dimensions of space that we experience is curved through a fourth dimension, this 3D space doubles back upon itself in just the same way that 'straight lines' drawn upon the surface of a globe double back upon themselves. [197]

The observation of the gravitational lens is proof that our universe is four–dimensional and it also proves that everything in it is likewise four–dimensional down to its atomic structure. Einstein did not come up with the Theory of Relativity on his own. A man named William Kingdom Clifford started work on the Theory of Relativity. Clifford laid the foundation of relativity, but he did not get very far beyond that because he passed away. Einstein added to the work that he started on the Theory of Relativity. William Kingdom Clifford expressed James Clerk Maxwell's electromagnetic theory using four – dimensional geometry. James Edward Biechier says that William Kingdom Clifford claimed that what we sense as matter is three – dimensional space curved in a fourth dimension. Clifford based this claim on his understanding and interpretation of Reinmann's geometry. Clifford also said that what we sense as matter in motion is no more than variations in a four–dimensional curvature. [198.1, 198.2]

Biechier tells us that: "Riemann's work directly implied that space is four dimensional as well as continuous." [199]

Biechier goes on to say that William Kingdom Clifford claimed that what we sense as matter is nothing more than our three spatial dimensional space curved into a fourth spatial dimension and what we conceive as matter in motion is nothing more than variations in a four-dimensional curvature. [200]

It was William Kingdom Clifford who came up with the four–dimensional curving of space that Einstein used to show how the warping of space can distort light's path through space – the gravitational lens. [201]

The General Theory of Relativity was proven when a gravitational lens was observed during a solar eclipse in 1919. [202]

This is exactly what Einstein said in his theory of relativity, that gravity is not a force at all. Gravity does not originate inside any atom. It has nothing to do with quantum mechanics, and it never will. As Einstein said: "This is a physical property of space – space acts on objects, but objects do not act on space." [203]

Larry D. Kirkpatrick and Gerald F. Wheeler said that if you invoke the equivalence principle that Einstein came up with, by equating acceleration with gravity, then you could safely conclude that gravity alters space. [204]

Kirkpatrick and Wheeler went on to say that, there is evidence for the distortion of space when light is bent by a gravitational field. For example, if there are three space stations in orbit around the Sun with astronauts in each of the stations. When astronauts in one space station takes sightings to obtain the angle between the other two space stations, they discover that the sum of the angles of the triangle formed by the three space stations is greater than 180^0, as illustrated in Figure 116. [205]

Kirkpatrick and Wheeler tell us that the astronauts on each space station are faced with a dilemma. Light is supposed follow a straight line in a vacuum. If light traveled a straight path, then the angles that all the astronauts in all the space stations would equal to 180^0. However, when the astronauts measure the angles between each space station all the angles add up to more than 180^0, because the space in which the space stations inhabit is curved through a fourth spatial dimension. [206]

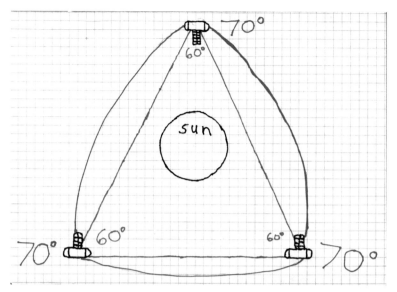

FIGURE 116

PART 3

EMBEDDED TIME TRAVEL

Einstein had problems with time paradoxes showing up in his General Theory of Relativity. Other astrophysicists, such as Stephen Hawking, have had similar problems with the time paradoxes in General Relativity. Stephen Hawking had even said that there must be some sort of Chronology Protection built into our universe because we do not see any of the paradoxes that General Relativity predicts. [207]

Let us look at these paradoxes and try to explain them. The first thing that you must understand is that we are **EMBEDDED** within the universe – we do not float above it, nor do we float under it. We know this because the only way to look through a gravitational lens is to be embedded in the same structure in which the lens resides.

Some astrophysicists think that our universe is like an inflating bubble. It is an inflating hyper bubble. Now we know that our hyper bubble universe cannot go backward in time – contract – unless there is a tremendous amount of heat applied to it. And just where would all that heat come from? Therefore, the past is closed to us on the grounds of simple thermodynamic laws. What about time travel paradoxes involving the future, are they possible?

The idea of traveling back through time by traveling faster than the speed of light is used in **Star Trek** episodes more than 22 times. [208]

The idea that you can travel back through time just by going faster than the speed of light is false. To see why this is the case let us do another little thought experiment. Let us say that you want to travel back through time to meet your great, great, great grandmother. Therefore, you get a Faster Than Light warp drive spaceship and set out on your journey to the past. You will notice that you and your spaceship are **EMBEDDED** within the four–dimensional structure of the universe. Therefore, when you set out on your journey as you travel faster than the speed of light in your space ship the fabric of the universe in which you are **EMBEDDED** travels, as it always does, into the future and dragging you and your space ship along with it. No matter how fast you travel, you can never travel back through time because the **space in which the space craft is EMBEDDED in is moving at a normal clip into the future.** Therefore, no matter how fast the **EMBEDDED** spaceship moves through the space in which it is **EMBEDDED**, the space itself never halts its normal movement through time into the future. The Figures 117 (key) and 118 illustrate this. It looks like Einstein was incorrect to say that **space** and **time** are the same, or that time is a dimension.

In the Figures 117 (key) and 118, you can see that the ship cannot move faster than the space that surrounds it. As space moves into the future so does the ship. The ship gets dragged along with the space in which it is **EMBEDDED**. The ship will always move into the future at the same speed as the space around it. The ship – being **EMBEDDED** within four–dimensional space – is prevented from moving into the future faster than the space it is enmeshed in and the ship is likewise totally prevented from moving into the past.

This means that the time paradox of "going back through time and killing your mother and father before you were born" is **PHYSICALLY** impossible because the **STRUCTURE** of the universe will not allow it.

The universe's expansion rate will always push you forward in time and space no matter how fast you travel through the universe.

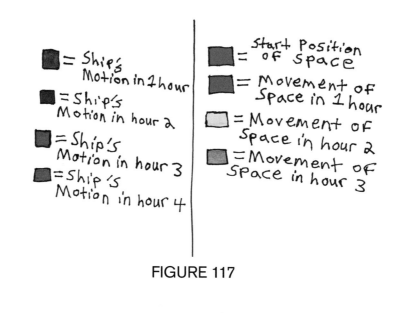

FIGURE 117

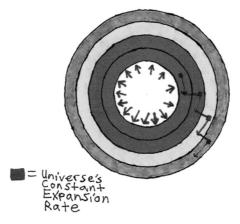

FIGURE 118

In Figure 119, we see how our universe is being pushed into the future by 30,000 g's of force and we also see how it would take 60,000 g's of force to push our universe backward through time.

Remember in the first section of this book, when we said that if we wanted the universe to contract into a Big Crunch then all we would have to do is heat it up. In this case – our universe is just like an ice cube, and like an ice cube just because you heat it up until it melts does not necessarily mean that the ice cube is going backward through time as it melts. Not only that but where would we get the 60,000 g's of force needed to condense our universe and how would we apply it uniformly to the outside of our universe?

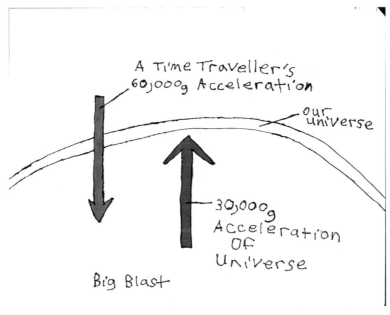

FIGURE 119

Can there ever be enough energy, force or pressure to compress our universe back into the stellar core from which it came? If so then our universe could return to its beginning initial conditions (see Figures 121 and 122 in Part 4). It would be a bouncing phoenix universe if it could return to its initial conditions.

PART 4

MOTION ORIENTED TIME TRAVEL

We have previously defined "time" as a motion in space that is relative to its surrounding environment. That being said, some motions are more reversible than others. One example of this can be seen with Laval Nozzles. One example of a Laval Nozzle is a rocket engine.

Donald Kingsbury poses a few interesting questions concerning time travel.

The first question he asks involves a sample of uranium 238 that we have in our lab. If we isolate an atom of it that has been stable for the billions of years that it has existed outside of its mother nova then is it guaranteed to be stable if we reverse time to return it back to its mother nova? In other words, can the uranium atom that we have in our lab return, through time, to its mother nova using the same path that it took to get to our lab? Kingsbury's answer to this question is no. [209]

Kingsbury then asks, if time is reversed will entropy decrease? Again, his answer is no. Kingsbury says that if time is reversed it would be difficult or impossible for rivers to slosh uphill and for the water in the ground water to rise up and levitate upward as raindrops before becoming rain clouds. [210]

In Figure 120 I depict the same problem. If a glass cup were to fall from a table and smash on the floor, then how do you reverse time so that all the pieces levitate from the floor and reform themselves into the original glass cup? In other words what force would allow this to happen?

FIGURE 120

In Figures 121 and 122 we see the thermodynamic evolution of our Observable Universe. Figures 121 and 122 show the Enthalpy of the Observable Universe. Before we look at Figures 121 and 122 in more detail let's discuss what Enthalpy means.

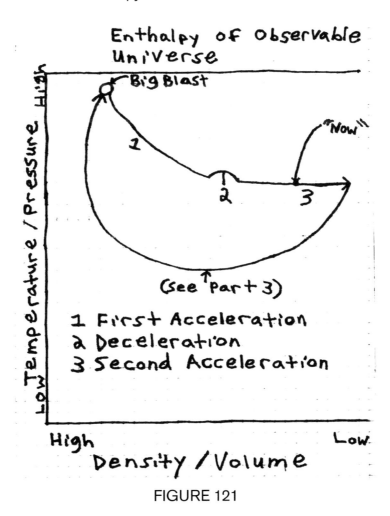

FIGURE 121

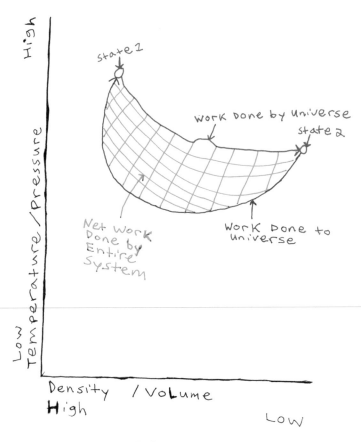

FIGURE 122

Enthalpy is a measurement of how much energy a thermodynamic system has. Enthalpy is the internal energy of the thermodynamic system plus the work (in terms of energy) that the thermodynamic system must do to create room for the thermodynamic system to exist in a given environment. A good analogy for what all of this means can be seen in the following thought experiment that Daniel Schroeder describes: [211]

Imagine a magician wanting to summon a rabbit out of nothing using the quantum fluctuations in a vacuum to help him in this task. Let's not look at the impossibility of the magician's task and instead let's concentrate on the Enthalpy of the rabbit that the magician is trying to summon. Let's build an equation as we examine the Enthalpy of the rabbit as the magician summons it from the quantum fluctuations in a vacuum. The first thing that the magician must do as he is summoning the rabbit out of the quantum fluctuations in a vacuum is to build it with all the matter and energy it will consist of (let's call this variable in the equation we are building "N"). In order to build the rabbit, the magician must expend energy in the form of work to construct the rabbit out of the quantum fluctuations in a vacuum (let's call this variable "P"). Now the magician must spend more energy to provide the rabbit with its internal energy in order to give it life (let's call this variable "Q"). Ok now the magician has everything that he needs to summon the rabbit, but first he must provide a place in the environment for the rabbit to exist in. For example, if the magician wants to summon the rabbit onto a table, the magician must first push away all the air (in the form of a rabbit) so that the rabbit can materialize in the vacuum thus created. This requires some work also (let's call this variable "W").

Now let's put all the variables into an equation – (N + P + Q + W = the total Enthalpy of the rabbit).

How would we use such an equation to calculate the Enthalpy of our universe? Let's do a comparison contrast between the Enthalpy of our universe and the Enthalpy of you. Remember in Section 1 Part 2 where we showed that the Pauli Exclusion Principle makes 'time travel' very difficult because some of the atoms in you now once belonged in your mother and grandmother and so on down though the first living cells that ever existed on Earth in an unbroken chain of life. If this were not so then you would not be here. With this in mind, then if we look at the Enthalpy in you does it mean that we have to look at the Enthalpy of all your ancestors as well? If so, then what about the Enthalpy of our universe? Would we have to look at the Enthalpy of all the mother universes that gave birth to our own universe? If our universe is part of a Self – Reproducing Eternal Inflation, then our universe' Enthalpy would be infinite. [212]

If you use a Cosmic Box to look at our universe's Enthalpy as it is now, then you won't have to worry about its origins as much.

Let's look at Figures 121 and 122. As we stated earlier, our Observable Universe started out as a hot dense iron stellar core that was ejected out of another universe. Once the stellar core reached the Bose – Einstein Condensate Hydrogen that surrounded its parent universe the stellar core detonated in an explosive decompression – Big Blast (Big Bang) event. We can designate this in the graphs of Figures 121 and 122 as State 1. I based the graphs of Figures 121 and 122 on a video from Khan Academy. [213]

Another way to look at this event is to look at it like this: when a White Dwarf star gains hydrogen from a neighboring star it starts to gain mass. When this happens the White Dwarf stellar remnant will try to restart its fires of nuclear fusion. Once the White Dwarf reaches 1.4 solar masses it becomes unstable and becomes a thermonuclear runaway. It then promptly explodes as a Type Ia Supernova.

If a neutron star were to steal hydrogen from a neighboring star the same thing would happen. So, it is not surprising that when a hot iron stellar core gets ejected from its parent universe into the Bose – Einstein Condensate Hydrogen that is outside its parent universe, the hot iron stellar core detonates in a spectacular explosion that not only fuses hydrogen into helium and lithium, but it also creates the conditions in which a new universe will form. One can consider our entire Observable Universe as a product of a stellar remnant that has come full circle in a long evolutionary process.

One possible way this process may happen can be seen in Figure 123. In the first part of Figure 123 a stellar core gets ejected from its mother universe into the surrounding Bose – Einstein Condensate Hydrogen medium. In the second part of Figure 123 while the stellar core is in this medium it briefly starts to gain mass and as it does so, it starts to fuse hydrogen into helium. The third part of Figure 123 depicts when the stellar core becomes a thermonuclear runaway, like a Type Ia supernova. The fourth part of Figure 123 is the resulting Big Blast (Big Bang) event. There is a problem with this type of possibility and that is that all supernova that we have ever seen produce large quantities of heavy elements such as Titanium, Calcium, Carbon, Oxygen, Silicon, Sulfur and Fe (iron). So, this hypothetical scenario is very unlikely.

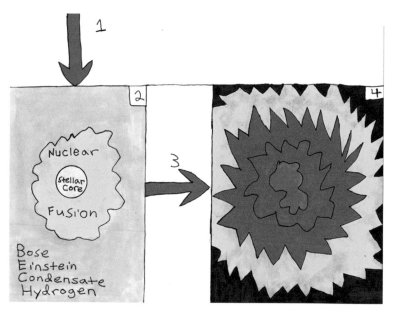

FIGURE 123

A more likely scenario would happen when a universe ejects a stellar core into the Bose – Einstein Condensate Hydrogen medium. When the stellar core is injected into this medium it is travelling faster than the speed of light and when it decelerates bellow a critical threshold it feels the effects of the extremely cold low-pressure environment that it has been ejected into.

Only neutron stellar cores get ejected out of universes, white dwarfs cannot be ejected out of their universe because the stars that produce white dwarves are not big enough to produce the thermal pressure needed to help eject a white dwarf stellar core out of its parent universe. White dwarf stellar cores are like neutron stellar cores. White dwarf stellar cores are held up by Electron Degeneracy Pressure and neutron stellar cores are held up by Neutron Degeneracy Pressure. Both types of degeneracy pressures are due to the Pauli Exclusion Principle. The Pauli Exclusion Principle states that no two electrons or neutrons can occupy the same place and have the same spin (angular momentum) at the same time. [214]

Neutron Degeneracy Pressure is so strong that nothing in any universe you could imagine can squeeze the neutron stellar core down any farther.

There are different types of neutron stars. For example, the neutron stars that are comprised mostly of hydrogen produce a sub luminous super nova explosion and the neutron stars that have a lot of iron in them produce a very luminous super nova (see Figures 124 and 125).

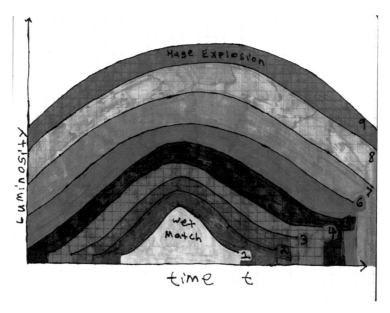

FIGURE124

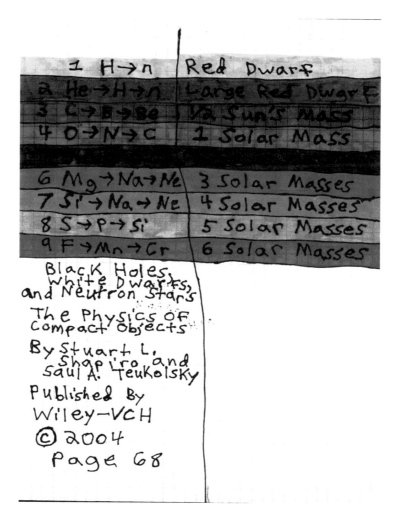

1 H→n	Red Dwarf
2 He→H→n	Large Red Dwarf
3 C→B→Be	1/2 Sun's Mass
4 O→N→C	1 Solar Mass
6 Mg→Na→Ne	3 Solar Masses
7 Si→Na→Ne	4 Solar Masses
8 S→P→Si	5 Solar Masses
9 F→Mn→Cr	6 Solar Masses

Black Holes,
White Dwarfs,
and Neutron Stars
The Physics of
Compact Objects
By Stuart L.
Shapiro and
Saul A. Teukolsky
Published By
Wiley-VCH
© 2004
Page 68

FIGURE 125

The top line in the graph in Figures 121 and 122 that slopes down represents the evolution of the universe that is formed in a Big Blast (Big Bang) event. This Big Blast event accelerates light (thermal radiation) and matter (hydrogen, helium, and lithium) at high speeds and this is represented by Line Segment 1. Once the momentum of the blast starts to dissipate everything starts to decelerate and this is represented by Line Segment 2. Then the Hyper Holes start to pump in new matter into the newly formed universe and when this happens a new phase of acceleration begins, and this is represented by Line Segment 3.

Edward Harrison says that the pressure at the edge of our Observable Universe does work to our Observable Universe. It produces energy – possibly mechanical energy. The Observable Universe as a whole, conserves energy in this way. [215]

Our Observable Universe also conserves energy in other ways. For example, Alex Filippenko tells us that the overall temperature of our Observable Universe went down faster since the Big Blast (Big Bang) event than it would have if there were no dark energy. Filippenko also states that as space expands in our Observable Universe light impedes the expansion. This means that light is doing work on the Observable Universe. As light is being redshifted by the expansion of our Observable Universe it loses energy and it is this lost energy that does work to our Observable Universe by impeding its expansion. [216]

Light warps the space in which our galaxy resides.

The initial conditions that our Observable Universe started in was a solid iron core being ejected into a sea of Bose – Einstein Condensate Hydrogen. This leads to a transition to a Big Blast Event that changed the solid iron stellar core and the Bose – Einstein Condensate Hydrogen into a hot plasma that was composed of hydrogen, helium, and lithium. As this plasma rapidly expanded it cooled. And as it cooled, heat as infrared light is dissipated to the Bose – Einstein Condensate Hydrogen surrounding it.

The matter that is our Observable Universe started in a solid state that was immersed in a Bose – Einstein Condensate state which then exploded. When this happened the solid and the Bose – Einstein Condensate did a phase transition to plasma. Parts of this plasma collected into small clumps forming stars and galaxies.

As we have seen throughout this book thermodynamic can have some unexpected connections to every aspect of physics. For example, Thermodynamics is related to sound.

After the Big Blast (Big Bang) event produced not just a lot of energy that produced the Cosmic Microwave Background Radiation (CMBR) that we see today it also produced a huge shock wave. A shock wave is a pressure wave and these waves are acoustic. [217]

There is a deep connection between thermal electromagnetic waves, sound waves, and gravity waves. As we have stated in Section 1, thermal electromagnetic waves (light) causes gravity waves (the warping of space) and they do this because they also produce pressure waves. This can be seen in the experiment in Section 1, Part 5.

Also, sound waves can (in certain situations) produce light through a process known as sonoluminescence. [218]

Another example of the deep connection between thermal electromagnetic waves, sound waves, and gravity waves can be seen in the Cosmic Microwave Background Radiation (CMBR). The Cosmic Microwave Background Radiation (CMBR) comes from a time when the temperature of our universe was 3,000 degrees Kelvin. All this energy got red shifted as our universe explosively expanded. [219]

The Cosmic Microwave Background Radiation (CMBR) is slightly patchy. It varies in brightness by 0.01 percent from one part of the sky to another due to the variations in Temperature (thermal electromagnetic radiation – light), density, and pressure from place to place. [220]

There has been a long history of studying the acoustics of the universe starting with Johann Kepler. [221]

The concept of "time" is a human invention that is based on things in nature (synchronous motion). There are natural objects and then there are objects, like cars, that only humans can produce. The beginning of "time" was when a human built the first "time" keeping device (the sundial). The human concept of "time" and the human invention of the clock that reinforces this concept are both based upon natural motions, like celestial motions or the resonances of quartz crystals and cesium atoms. The human concept of music is also based on the motion and resonances of objects found in nature. A fallen hollowed out tree would make a great drum, for instance. A vibrating drum head may have inspired Alan Guth to develop his theory of inflation. In Figure 126 we see how a drum head warps as it vibrates after it has been struck and then in Figure 127, we see a picture depicting Alan Guth's theory of inflation.

Throughout this book we have seen how thermodynamics has shaped and molded our universe into what we see today.

The effect of this scientific knowledge is manifested in the technologies that come from such knowledge and it is these applied sciences that we will look at in the Third Section of this book.

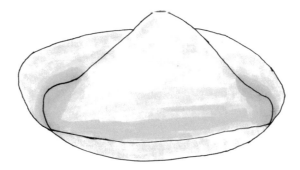

FIGURE 126

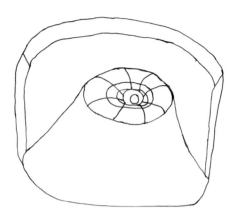

FIGURE 127

ROGER I. PARKER II

SECTION 3

TECHNOLOGY APPLIED SCIENCE

PART 1

WARP DRIVE

What is 'science'? At its core science is humanity's basic understanding of how nature – the universe works. What is 'technology'? Technology is when humanity uses our understanding of how the universe – nature works in order to better our lives in some way. This section will look at the Applied Science – the technology that will emerge from my work.

Now that we have proven that there are no quantum fluctuations in a vacuum with the first experiment in this book, we can now look at what it means for space travel. In Section 1, Part 4 of this book I talk about the debate as to what the Casimir Effect really is.

In 1994, Miguel Alcubierre gave a mathematical theory of Warp Drive that is consistent with Einstein's General Theory of Relativity. Hubble discovered that the space that makes up our universe is expanding faster than the speed of light. [222]

Miguel Alcubierre used this loophole in Relativity's claim that nothing can go faster than the speed of light to formulate his mathematical theory of Warp Drive. He thought that if the ship could ride a spatial wave then the ship would theoretically be able to travel faster than light because the ship would not be traveling **through** space, instead the ship would be traveling **with** space as the space moved. [223]

As we have seen in Sections 1 and 2 of this book thermal fluctuations causes space to warp. Cold Void Spikes are gravitational mountains and hot galactic superclusters are lush valleys. The warp drive starship is like a sled using gravity to slide down a mountain slope to the valley below (see Figure 130).

In Figure 131 I recreated a picture that Miguel Alcubierre used to represent a warp field. [224]

Warp drive works by compressing space in front of the spaceship and expanding the space behind the spaceship. In the first section of this book we found that to compress space you had to heat it and that to expand space you had to cool it. So that means that one way to have a working warp drive would be to heat up the space in front of the spaceship to compress it as you cool the space behind the spaceship to expand it in order to create a warp bubble and to propel the ship at warp speeds (see Figure 128 and Figure 129).

The consequence of creating a warp bubble in this way is that the whole spacecraft will be lifted into four-dimensional space (we talked about four-dimensional space in section 2 of this book).

NASA has been working on warp drive for a while now and has some interesting ship designs. Figures 128 and 129 are based on one such design. [225]

Mae Jemison started the 100 Year Starship initiative. The 100 Year Starship initiative is designed to make space travel to another star system possible in 100 years. [226]

As you can see in the first warp drive design (Figures 128 and 129) you have a heating unit and a cooling unit. The heating unit heats up the space in front of the starship while at the same time the cooling unit cools the space behind the starship and in this way a warp bubble is produced that propels the ship forward.

Will this design work? The answer to that question is unknown. In theory it should work.

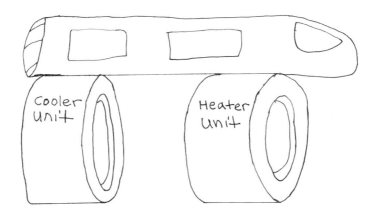

FIGURE 128

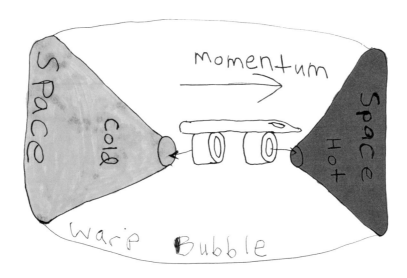

FIGURE 129

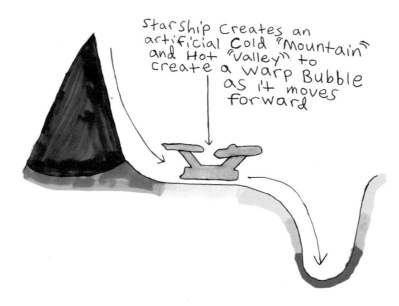

FIGURE 130

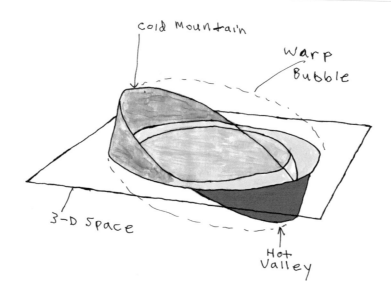

FIGURE 131

Another way a warp drive unit might be able to work is depicted in Figures 132, 133, 134, and 135. In this warp drive design, you have a warp nacelle. The warp drive nacelle has a reaction chamber in the front. A vacuum is maintained in this reaction chamber that is heated to very high temperatures then instantly cooled to very low temperatures. This produces thermal shock waves as well as gravity waves that are then injected into a long tube. By doing this it compresses the thermal shock waves and the gravity waves. While the thermal shock waves and the gravity waves are in the tube they are in a state of maximum scrunch. This means that when they leave the tube that extends to the end of the warp drive nacelle the compressed thermal shock waves and gravity waves will quickly expand once they leave the confines of the nacelle, thus propelling the space craft forward.

Thermal fluctuations in a vacuum can warp space itself and we have proven that experimentally earlier in Section 1, Part 5 of this book. The warped space that is produced by some thermal reactions in a vacuum are known as gravity waves. We might be able to produce and use these gravity waves to produce and maintain a warp bubble around a starship. The starship would ride a shock wave. As can be seen in Figures 132, 133, 134, and 135; thermal pressure waves and gravity waves are produced in the reaction chamber of the warp nacelle and then they are forced out the back of the nacelle. When the thermal pressure waves and the gravitational waves leave the nacelle, they quickly expand thus pushing the starship forward along thermal and gravitational pressure waves. The thermal and gravitational pressure waves when they leave the nacelle they are squashed (like a coiled spring) they are at a state of maximum scrunch so when they are released from the nacelle they recoil and quickly expand and as they do so they push the starship forward.

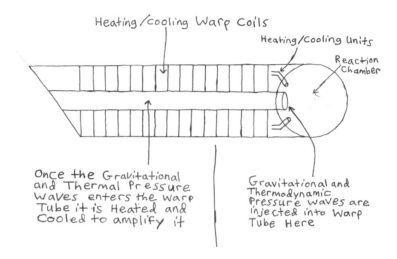

FIGURE 132

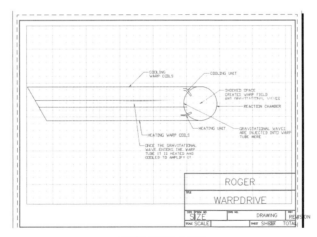

FIGURE 133

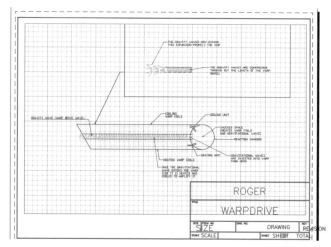

FIGURE 134

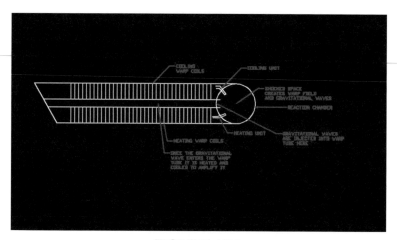

FIGURE 135

Miguel Alcubierre developed a theory of how a warp drive would work in 1994. His warp drive theory does not violate Einstein's General Theory of Relativity, which states that nothing can travel faster than the speed of light. There is no time dilation with his theory either. [227].

There are two ways of achieving Faster Than Light space travel that does not violate Einstein's Theory of Relativity. One way is to create a "wormhole" – (see section 2: Part 2: Proof 2: Hyper Holes) – and the other way is to create a "warp bubble" around the space craft that propels it at Faster Than Light speeds.

The 1960's television show **Star Trek** set lofty technical and social goals into the future. We have achieved some of the technical goals, but we still have a long way to go to achieving all the lofty goals that **Star Trek** set for us. [228]

The designs depicted here are way ahead of their time like Leonardo Da Vinci's design of a helicopter. His basic idea that if you spin a propeller fast enough you could get a craft off the ground was correct. However, there was no power available to Da Vinci to make his helicopter work. He lived hundreds of years before the steam engines and the internal combustion engines of the Industrial Revolution. Plus, Da Vinci's propeller design was wrong. (See Figure 136.)

The basic idea that by heating and cooling space (a vacuum) you can create a warp bubble to propel a space craft is correct enough, but like Da Vinci, the two designs that I have given here are probably wrong. For example, I do not know how to get the power needed to heat then rapidly cool space.

Also, there is a slight problem with any warp drive engine that I have depicted here. And that is if you built either of the warp drive designs that I discussed on Earth and tried to launch them from the Earth's surface neither of the designs would get off the ground. Both designs CAN ONLY BE USED IN SPACE. Both designs can only work in the vacuum of space.

In the next section we will discuss in more detail about how four-dimensional communication is technically possible.

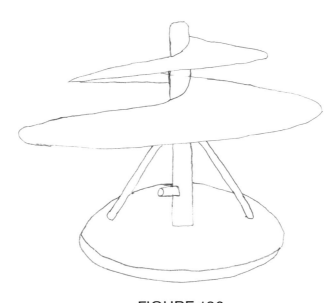

FIGURE 136

PART 2

FOUR-DIMENSIONAL FASTER-THAN-LIGHT COMMUNICATIONS

In Section 2, Proof 4 we discussed how four-dimensional communication is theoretically possible now we will discuss how four-dimensional communication is technically possible.

How would we make four-dimensional communication possible? In order to answer that question, let us first look at four-dimensional hypercubic crystals. These crystals show us that everything in our universe is four-dimensional down to its atomic structure and they show how higher dimensional objects cast lower dimensional "shadows". Light shining on a transparent four-dimensional object will cast a three-dimensional "shadow" just as a light shining on a transparent three-dimensional object will cast a two-dimensional shadow. (See Figure 113.)

Let us do a thought experiment where we have a hypercubic quartz crystal. If light can be cast from a higher dimension to a lower one (thus forming a "shadow") then we might be able to cast light from a lower dimension to a higher dimension also. For example, if quartz crystals are hypercubic then that means that we could send a photo-electromagnetic pulse though it and expect it to propagate through the fourth dimension. If we transmitted a radio signal in this way, then we would be a step closer to four-dimensional communication. Stan Gibilisco talks about such a possibility (see Figure 137). [229]

Radio waves are also very useful in navigation. Radar uses radio frequencies to track the movement of aircraft, for example, and four – dimensional radar can be used to track the movement of starships as they move through four-dimensional space. (See Figure 138)

In order for our scientific knowledge and technologies to develop further, multi-disciplinary science must be taught in all schools. [230]

Another way that four-dimensional crystals will be increasing our technical capabilities is by allowing us to create new materials. For example, a certain type of bismuth and palladium crystal alloy forms Mobius strip rings that allow an electrical current to cycle clockwise and counterclockwise simultaneously. And in this way it can be a quibit. A bit exists in one of two stated 1 (on) or 0 (off), but a quibit can exist in two states at once. The 1 (on) state and the 0 (off) state can exist together at the same time in a superposition of both states because the Mobius strip ring allows a single current that enters the

Mobius strip ring to be on and off because you can see both sides of a Mobius strip ring at once (see Figures 139 and 140). This will be a huge step forward in the development of quantum computers. [231]

With better, more accurate knowledge the way we view the universe changes. This allows us to develop better technology that changes our lives for the better as it also changes how we view the universe.

There has been a debate for a long time as to whether a photon of light is a particle or a wave. Some experiments show that a photon is a particle while other experiments show that a photon is a wave and still other experiments show that a photon is both a particle and a wave. [232]

Let us explore this controversy, remember in Section 1, Part 3 we learned that some physicists have assumed that a photon has "no mass". One of the reasons that they believe this to be true is that they were using the wrong equation to calculate the mass and energy of the photon. [233]

Newton's First Law of Motion (Inertia) says that an object at rest tends to stay at rest unless it is affected by an outside net force. Newton's First Law of Motion (Inertia) also says that an object in motion tends to stay in motion unless it is affected by an outside net force. [234]

What we learned in Section 1, Part 3 is that nothing in our universe is at rest (static). Everything in our universe is in some type of motion from the vibration of the atoms that make up an object to the cosmic movements of stars and planets it is all in motion. Therefore, when we consider whether something is at rest or motion, we are really looking at whether the object appears to be at rest or motion in a particular Frames of Reference.

For example, if we had two boats — let us call them boat A and boat B. Now we can do a thought experiment where we have boat A travel at one mile per hour in an ocean current and boat B is traveling at 132 miles per hour. If boat A and boat B were in the same part of the ocean and if boat B were speeding to boat A at 132 miles per hour then it would appear to anyone on boat B that boat A is motionless relative to boat B's Frame of Reference.

Let us now look at the mathematics needed to calculate a photon's mass and energy. The equation needed to calculate the Rest Energy and Rest Mass of an object is $E=MC^2$.

As we have stated earlier, nothing in our universe can be considered "at rest" if we define "at rest" as motionless — everything in our universe is in motion so this definition of "at rest" no longer works well. However, Newton's second definition of "at rest" -- that an object in motion tends to stay in motion unless it is affected by an outside net force — does work and is a more accurate description of what "at rest" really means from a cosmic perspective. When calculating the mass and energy of a photon this equation should be used. However, some physicists interpreted "at rest" to mean immobile and as we have seen this is incorrect.

Some physicists use the equation used to calculate an object's kinetic energy when they calculate the mass and energy of a photon. This is the incorrect equation to use.

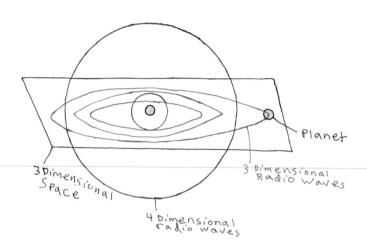

3 Dimensional Space

3 Dimensional Radio Waves

Planet

4 Dimensional radio waves

FIGURE 137

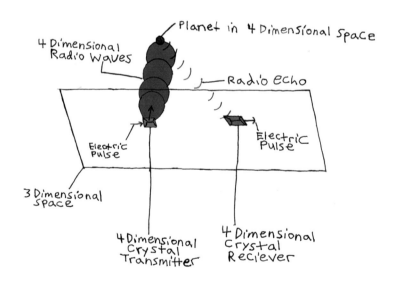

Planet in 4 Dimensional Space

4 Dimensional Radio Waves

Radio Echo

Electric Pulse

Electric Pulse

3 Dimensional Space

4 Dimensional Crystal Transmitter

4 Dimensional Crystal Reciever

FIGURE 138

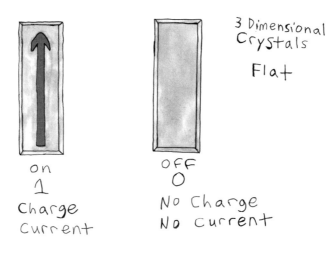

on
1
Charge
Current

OFF
0
No Charge
No Current

3 Dimensional
Crystals

Flat

FIGURE 139

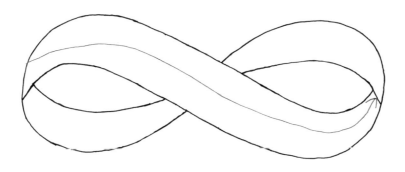

FIGURE 140

BIBLIOGRAPHY

INTRODUCTION

1) Bertrand Toudic, Pilar Garcia, Christophe Odin, Philippe Rabiller, Claude Ecolivet, Eric Collet, Philippe Bourges, Garry J. McIntry, Mark D. Hollingsworth, and Tomasz Breczewski, Hidden Degrees of Freedom in Aperiodic Materials, **Science** January 4, 2008, pages 11, 41–42, 69–71, https://science.sciencemag.org/content/319/5859/69

2) John Langone, **National Geographic's How Things Work**, published by National Geographic, © 1999, page 41

3) Carl Sagan, **Cosmos**, Episode 10: The Edge of Forever, © 1980

4.1) Rudy Rucker, **The Fourth Dimension**, published by Houghton Mifflin Company, © 1985, pages 82–83

4.2) Edward R. Harrison, **Cosmology**, published by Cambridge University Press, © 1981, pages 154–155

5.1) James Edward Biechlier, *Three Logical Proofs: The Five – Dimensional Reality of Space – Time*, **Journal of Scientific Exploration**, Vol 21, No 3, © 2007 page 523 – 542 https://pdfs.semanticscholar.org/ee33/5fbf33ba621e189fc5452ba4438f85fdabfa.pdf

5.2) Wikipedia contributors – primary contributor Revision History Statistics, **William Kingdom Clifford,** published by Wikipedia Free Encyclopedia, at https://en.wikipedia.org/w/index.php?title=Wiliam_Kingdom_Clifford&oldid=826789941, May 21, 2017, pages 1 – 9

5.3) Edited by William E. Baylis, **Clifford Geometric Algebras**, published by Birkhauser, © 1996, pages 237 – 252

6) William Kingdon Clifford, **On The Space Theory of Matter**, © 1876

7) Tucker Hiatt, **How Fast Are You Moving?**, TED Ed, © 2014, https://www.youtube.com/watch?v=wlzvfki5ozU

SECTION 1
PART 1: PROBLEMS WITH MODERN PHYSICS

8)   Roger Penrose, ***Fashion, Faith and Fantasy: In the New Physics of the Universe***, published by Princeton University Press, © 2016, pages xi – xii

9)   Neil deGrasse Tyson, ***Adventures in Science Literacy Part 1 and Part 2***, University of Wisconsin – Madison, © 2009, http://youtu.be/ClNWybVx5c /

10) Tim Folger, ***Discover***, May 2017, The War Over Reality, pages 28 – 33

11) Tim Folger, ***Discover***, May 2017, The War Over Reality, pages 28 – 33

12) Roger Penrose, ***The Emperor's New Mind***, published by Penguin Books, © 1991

13.1) Neil deGrasse Tyson, ***Cosmos: A Spacetime Odyssey***, Episode 13 Unafraid of the Dark, © 2014

13.2) interview done by Tom Bilyeu for Motivation Hub's interview with Neil deGrasse Tyson, June 3, 2020, https://www.youtube.com/watch?v=JtahB1-MNvk&t=37s

14) Larry King interview with Neil deGrasse Tyson, July 12, 2018, https://www.youtube.com/watch?v=5mDzcxy2KVI&t=15s

15) Massimo Polidoro, ***Skeptical Inquirer***, May / June 2019, The Mind of Leonardo da Vinci, Part 2, pages 23 – 24

PART 2: PAULI EXCLUSION PRINCIPLE

16.1) Joanna Baker, ***50 Physics Ideas: You Really Need to Know***, published by Quercus, © 2007, pages 120—123

16.2) Kitty Ferguson, ***Prisons of Light: Black Holes***, published by Cambridge University Press, © 1996, page 14

17) Paul Francis, ***A3 L2 VD4 Degeneracy Pressure***, July 30, 2014, https://www.youtube.com/watch?v=adHBd1y8yek

18) Kenneth Libbrecht and Patricia Rasmussen, ***The Snowflake: Winter's Secret Beauty***, published by Voyager Press, © 2003, page 45

19.1) Carl Sagan, ***Cosmos***, published by Random House, © 1980, pages 265 – 267

19.2) Jean – Fancois Gouyet, ***Physics and Fractal Structures***, published by Springer, © 1996, pages 41 – 45

19.3) John D. Barrow, ***The Book of Universes***, published by W. W. Norton & Company, © 2011, page 87

19.4) P. J. E. Peebles, ***Principles of Physical Cosmology***, Published by Princeton University Press, © 1993, pages 209 – 224

20) Edward R. Harrison, ***Cosmology***, published by Cambridge University Press, © 1981, pages 83—87

21) Carl Sagan, ***Cosmos***, Episode 10: The Edge of Forever, © 1980

22.1) Harry A. Schmitz, ***Journal of Theoretics***: On the Role of the Fractal Cosmos in the Birth and Origins of Universes, © 2002, pages 1—13, http://www.journaloftheoretics.com/Links/Papers/Schmitz.pdf

22.2) John D. Barrow, ***The Book of Universes***, published by W. W. Norton & Company, © 2011, pages 87, 247

22.3) P. J. E. Peebles, ***Principles of Physical Cosmology***, published by Princeton University Press, © 1993, pages 209 – 224

22.4) Blair Macdonald, ***Fractal Geometry a Possible Explanation to the Accelerating Expansion of the Universe and Other Standard Lambda CDM Model Anomalies***, published by Academia, © 2014, pages 1 – 30, https://independent.academia.edu/BlairMacdonald or https://www.academia.edu/8415112/Fractal_Geometry_a_Possible_Explanation_to_the_Accelerating_Expansion_of_the_Universe_and_Other_Standard_%CE%9BCDM_Model_Anomalies

22.5) Blair Macdonald, ***Observed Galaxy Distribution Transition With Increasing Redshift a Property of the Fractal***, published by Academia, © 2014, pages 1 – 13, https://independent.academia.edu/BlairMacdonald or https://www.academia.edu/9412115/Observed_Galaxy_Distribution_Transition_with_Increasing_Redshift_a_Property_of_the_Fractal

23) J. Richard Gott, ***Time Travel in Einstein's Universe***, published by Houghton Mifflin Company, © 2001, page 139 – 140

24) Stephen Hawking, ***The Universe in a Nutshell***, published by Bantam Books, © 2001, pages 153, 160

PART 3: THE PHOTON'S MASS

25.1) Liang—Cheng Tu, Jun Luo, and George T. Gillies, The Mass of the Photon, Institute of Physics Publishing, ***Reports on Progress in Physics***, © 2004, pages 77 – 130, http://www.optica.machorro.net/Lecturas/PhotonMass_rpp5_1_RO2.pdf

25.2) G. Spavieri and M. Rodriguez, Photon Mass and Quantum Effects of the Aharonov-Bohm Type, arXiv:0705.1101[quant—ph] 8 May 2007, © 2007, pages 1 – 5, https://arxiv.org/pdf/0705.1101.pdf

26.1) Joanna Baker, **50 Physics Ideas: You Really Need To Know**, published by Quercus, © 2007, pages 20 – 23

26.2) Robert L. Lehman, **Barron's E – Z Physics**, © 2009, pages 410 – 412

26.3) Konrad B. Krauskopf and Arthur Beiser, **Fundamentals of Physical Science**, published by McGraw – Hill Books, © 1971, pages 266 – 267

26.4) Roger Penrose, **Fashion, Faith and Fantasy: In the New Physics of the Universe**, Published by Princeton University Press, © 2016, pages 55 – 56

26.5) Mr. Anderson, Bozeman science.com, **Mass–Energy Equivalence, Physics Essentials 089**, March 15, 2015, http://www.bozemanscience.com/ap-phys-089-mass-equivalent/

26.6) Stan Gibilisco, **Understanding Einstein's Theories of Relativity**, published by Dover Press, © 1983, page 16 – 18

26.7) Ricardo Gobato, Alekssander Gobato, and Desire Francine Gobato Fedrigo, **Energy and matter .02892v1 [physics.pop-ph] 2 Sep 2015**, © 2015, pages 1 – 15, https://arxiv.org/pdf/1509.02892.pdf

27) Brian Greene, **The Elegant Universe**, published by Vintage Books a Division of Random House, © 2003, page 11

28) Robert Kaplan, **The Nothing That Is**, published by Oxford University Press, © 1999, page 179

29) Sadri Hassani, **Skeptical Inquirer**, July / August 2019, Massless is Not Nonmaterial, pages 14 – 16

30) Hideo Nitta, Masafumi Yamamoto, and Keita Takatsu, **The Manga Guide to Relativity**, published by No Starch Press, © 2011, page 113

31) Udacity, **Godel's Incompleteness Theorem - Intro to Theoretical Computer Science**, February 23, 2015, https://www.youtube.com/watch?v=la6BK5X2LI8

32) Liang–Cheng Tu, Jun Luo, and George T. Gillies, The Mass of the Photon, Institute of Physics Publishing, **Reports on Progress in Physics**, © 2004, pages 77 – 130, http://www.optica.machorro.net/Lecturas/PhotonMass_rpp5_1_RO2.pdf

33) Miguel Alcubierre, The Warp Drive: Hyper–Fast Travel Within General Relativity, **Classic and Quantum Gravity**, © 1994, pages 73 – 77, https://arxiv.org/pdf/gr-qc/0009013.pdf

34) Gary Johnstone and David Bodanis, **Einstein's Big Idea**, Nova, © 2005

35.1) Stan Gibilisco, **Understanding Einstein's Theories of Relativity**, published by Dover Press, © 1983, page 84 – 86

35.2)   David Freedman, **Discover**, February 1989, Beyond Einstein, pages 56 – 61

36)   Miguel Alcubierre, The Warp Drive: Hyper–Fast Travel Within General Relativity, **Classic and Quantum Gravity**, © 1994, pages 73 – 77, https://arxiv.org/pdf/gr-qc/0009013.pdf

37.1)   Andrew Frankoni, How Fast Are You Moving When Your Sitting Still? **The Universe in the Classroom**, www.astrosociety.org/uitc, No. 71 spring 2007, pages 1 – 7

37.2)   Bob Berman, **Astronomy**, January 2012, Strange Universe: Poetry of Motion, page 12

37.3)   Tucker Hiatt, **How Fast Are You Moving Right Now**, TED Ed, © 2014, https://www.youtube.com/watch?v=wIzvfki5ozU

38)   Larry King Interview with Neil deGrasse Tyson, October 10, 2013, https://www.youtube.com/watch?v=C7kubIYu69c

39)   Hideo Nitta and Keita Takatsu, **The Manga Guide to Physics**, published by No Starch Press, © 2009, pages 70 – 74, 90 – 97, 112 – 115, 119, 139 – 140

40)   Joseph Schwartz and Michael McGuinness, **Einstein for Beginners**, published by Pantheon Press, © 1979, pages 155 – 163

41)   Hideo Nitta and Keita Takatsu, **The Manga Guide to Physics**, published by No Starch Press, © 2009, pages 118, 139 – 140, 144 – 146, 159

42)   Colin Pask, **Magnificent Principia**, published by Prometheus Books, © 2013, pages 100 – 101

43)   John C. Hodge, **Theory of Everything**, published by Blue Ridge Community College, February 18, 2012, pages 19 – 46

44)   James Edward Biechier, **2008 Anomalous Physics**, © 2008, https://www.youtube.com/watch?v=cnjpDuPkEWA

45)   Michel van Biezen, **Particle Physics (20 of 41)**, What is a Photon, May 20, 2015, https://www.youtube.com/watch?v=57pU2F-bIQs

46)   Michael Guillen, **Five Equations That Changed the World**, published by Hyperion, © 1995, page 253

47.1)   Orsola De Marco, W Schmutz, P A Crowther, D. J. Hillier, L. Dessart, A. de Koter, and J. Schweickhardt, The Gamma Velorum Binary System, **Astronomy and Astrophysics**, https://arxiv.org/pdf/astro-ph/0004081.pdf or **arXiv:astro-ph/0004081v1**, pages 1 – 17, April 6, 2000

47.2)   David Bodanis, **E=MC²**, published by Walker and Company, © 2000, page 238

47.3)   Jeremy Bernstein, ***Einstein,*** published by Penguin Books, © 1978, page 98

48.1)   Robert L. Lehman, **Barron's E – Z Physics**, published by Barron's Educational Series, © 2009, pages 410 – 412

48.2)   Konrad B. Krauskopf and Arthur Beiser, **Fundamentals of Physical Science**, published by McGraw – Hill Books, © 1971, pages 266 – 267

48.3)   Roger Penrose, **Fashion, Faith and, Fantasy: In the New Physics of the Universe**, Published by Princeton University Press, © 2016, pages 55 – 56

48.4)   Stan Gibilisco, **Understanding Einstein's Theories of Relativity**, published by Dover Press, © 1983, page 16 – 18

48.5)   Henning Genz, **Nothingness**, published by Helix Books, © 1999, page 190

49)   Roger Penrose, **The Emperors New Mind**, published by Penguin Books, © 1989, pages 220, 224

PART 4: THE CASIMIR EFFECT

50)   Joanna Baker, **50 Physics Ideas: You Really Need to Know**, published by Quercus, © 2007, pages 20 – 23

51)   John C Hodge, **Theory of Everything**, published by Blue Ridge Community College, February 18, 2012, pages 19 – 46

52)   E. Elizalde and A. Romeo, Essentials of the Casimir Effect and its Computation, published by **American Journal of Physics**, August 1991, https://aapt.scitation.org/doi/abs/10.1119/1.16749

53)   Hamish Johnson, Thermal Casimir Force Seen for the First Time, **Institute of Physic: Physics World.Com**, February 8, 2011, pages 1 and 2, pages 711 – 719, https://physicsworld.com/a/thermal-casimir-force-seen-for-the-first-time/

54.1)   Jack Sarfatti, Casimir Force: The Irrelevant answer for the Wrong Reason? Hunt for Zero Point Energy: **Excerpt from Super Cosmos**, October 11, 2004, pages 1 – 20, http://zpenergy.com/downloads/PuthoffCook.pdf

54.2)   Wikipedia contributors – primary contributor Revision History Statistics, **Quantum Vacuum Thruster**, published by Wikipedia The Free Encyclopedia, April 23, 2017, pages 1 – 12, https://en.wikipedia.org/wiki/Quantum_vacuum_thruster or https://en.wikipedia.org/w/index.php?title=Quantum_vacuum_thruster&oldid=843550601

55)   Roger Penrose, **Fashion, Faith and Fantasy: In the New Physics of the Universe**, published by Princeton University Press, © 2016, page 287

56) Robert L. Forward and Joel Davis, ***Mirror Matter: Pioneering Antimatter Physics***, published by Wiley Science Editions, © 1988, pages 28 – 29

57) Brian Greene, ***The Fabric of the Cosmos Nova Transcripts "Universe or Multiverse?"*** page 4, (http://www.pbs.org/wgbh/nova/physics/fabric-of-cosmos.html) © 2011

58) Joanne Baker, ***50 Physics Ideas: You Really Need to Know***, published by Quercus, © 2007, pages 36 – 39

59) Joanne Baker, ***50 Physics Ideas: You Really Need to Know***, published by Quercus, © 2007, pages 52 – 55

60) Adam Hadhazy, ***Discover***, December 2016, "Nothing Really Matters", pages 46 – 53

61) Istvan Szapudi, ***Scientific American***, August 2016, "Emptiest Place in Space", pages 28 – 35

62) Robert E. Krebs, ***Scientific Laws, Principles, and Theories: A Reference Guide***, published by Greenwood Press, © 2001, page 186

63) Jennifer Johnson, ***American Scientist***, September /October 2018, "A Chemical History of the Universe", pages 264 – 265

64.1) John P. Cise, KINEMATICS Units 4 & 5, ***Cosmos Controversy: The Universe Is Expanding, but How Fast?*** published by Austin Community College, jpcise@austincc.edu, February 20, 2017, http://cisephysics.homestead.com/files/ExpandingUniverse.pdf

64.2) T. Wayne, MRWAYNESCLASS.COM, ***G's Felt In A Loop Example***, © 2013, https://www.youtube.com/watch?v=qE4Teq5na6U

64.3) Govert Schilling, Constant Controversy, ***Sky and Telescope,*** June 2019, pages 22 – 29

65) John Langone, Bruce Stutz, and Andrea Gianopoulos, ***Theories for Everything***, published by National Geographic, © 2006, page 211

66.1) P. Birch, Is the Universe Rotating? ***Nature Vol 298 No 5873***, July 29, 1982, pages 451 – 454, https://www.orionsarm.com/fm_store/IsTheUniverseRotating.pdf

66.2) Shi–Chun Su, M.–C. Chu, Is the Universe Rotating? © June 24, 2009, pages 1 – 20, ***Astrophysics Cosmology and Nongalactic Astrophysics, arXiv:0902.4575***

66.3) John D. Barrow, ***The Book of Universes***, published by W. W. Norton & Company, © 2011, pages 115 – 122

66.4) Evangelos Chaliasos, The Rotating and Accelerating Universe, ***arX:astro-ph 061659 27 Apr 2009***, page 1 – 24, https://arxiv.org/ftp/astro-ph/papers/0601/0601659.pdf

66.5) Clara Moskowitz, The Inner Lives of Neutron Stars, **Scientific American**, March 2019, pages 24 – 29

67) Richard L. Faber, **Differential Geometry and Relativity Theory an Introduction**, published by Marcel Dekker, © 1983, section 3.2, pages 172 – 181

68) Steven Holzner, **Physics I For Dummies 2nd edition**, published by Wiley Publishing Inc., © 2011, pages 355 – 356

69) Alex Filippenko, Dark Energy and the Runaway Universe, November 14, 2013, **Talks at Google**, https://www.youtube.com/watch?v=Guvv5olLxCQ

70) Alex Filippenko, Dark Energy and the Runaway Universe, November 14, 2013, **Talks at Google**, https://www.youtube.com/watch?v=Guvv5olLxCQ

71) Hamish Johnson, Thermal Casimir Force Seen for the First Time, **Institute of Physic: Physics World.Com**, February 8, 2011, https://physicsworld.com/a/thermal-casimir-force-seen-for-the-first-time/

72) Hendrik Casimir, On the Attraction Between Two Perfectly Conducting Plates, **Proceedings of the Royal Netherlands Academy of Arts and Sciences 51, (1948)**, © 1948, pages 61 – 63, http://www.digitallibrary.nl

73.1) **The Audiopedia**, Radiation Pressure: Meaning, Definition, Explanation, January 17, 2017, https://www.youtube.com/watch?v=DDJXFSwMCTw&t=152s

73.2) **AK LECTURES**, Electromagnetic Radiation Pressure, January 11, 2014, https://www.youtube.com/watch?v=1ZEmv2uM2Yc

74) Steven Holzner, **Physics I For Dummies 2nd edition**, Wiley Publishing Inc, © 2011, pages 356 – 357

75) Joseph Cugnon, The Casimir Effect, **Acta Physica Polonica B**, January 27, 2010, published by Jagelloninan University Institute of Physics, pages 539 -- 546, http://hdl.handle.net/2268/82727

76) Joseph Silk, **The Big Bang Third Edition**, Published by W. H. Freeman and Company, © 2001, pages 387 – 388

77) Adam G. Riess and Michael S. Turner, **Scientific American**, February 2004, From Slowdown to Speedup, pages 62 – 67

78) Adam G. Riess and Michael S. Turner, **Scientific American**, February 2004, From Slowdown to Speedup, pages 62 – 67

79) Adam G. Riess and Michael S. Turner, **Scientific American**, February 2004, From Slowdown to Speedup, pages 62 – 67

PART 5: THE MODIFIED EXPERIMENT

PART 6: CONCLUSION

80) Joanna Baker, **50 Physics Ideas: You Really Need to Know**, published by Quercus, © 2007, pages 96 – 98

81.1) Charles Osgood, **A Science Odyssey: Mysteries of the Universe**, © 1998, https://www.youtube.com/watch?v=Jn7VcOU3x2g&index=3&list=PLoJC20gNfC2hFT1oLqhuRfCBdsCaNU9pb&t=0s

81.2) Bill Bryson, * **A Short History of Nearly Everything**, published by Broadway Books, © 2003, pages 140–141

82.1) Charles Osgood, **A Science Odyssey: Mysteries of the Universe**, © 1998, https://www.youtube.com/watch?v=Jn7VcOU3x2g&index=3&list=PLoJC20gNfC2hFT1oLqhuRfCBdsCaNU9pb&t=0s

82.2) Bill Bryson, * **A Short History of Nearly Everything**, Broadway Books, © 2003, pages 140–141

83) John D. Barrow, **The Book of Universes**, published by W. W. Norton & Company, © 2011, pages 190

84) Amit Goswami, Richard Reed, and Maggie Goswami, **The Self–Aware Universe**, published by Jeremy P Tarcher / Putnam a member of Penguin Putnam Inc. © 1995, page 141

85.1) Joanna Baker, **50 Physics Ideas: You Really Need to Know**, published by Quercus, © 2007, pages 200 – 203

85.2) Stephen Hawking, **The Universe in a Nutshell**, published by Bantam Books, © 2001, pages 85 – 87

86) Edward R. Harrison, **Cosmology**, published by Cambridge University Press, © 1981, pages 13 – 18

87) Matt Parker, **Things to Make and Do in the Fourth Dimension**, published by Farrar, Straus and Gioux, © 2014, page 422

88) Roger Penrose, **Discover**, September 2009, Discover Interview: Roger Penrose Says Physics is Wrong from String Theory to Quantum Mechanics, pages 56 – 57

89) Matt Parker, **Things to Make and Do in the Fourth Dimension**, published by Farrar, Straus and Gioux, © 2014, page 422

90) James Edward Biechlier, **2012 Vigier8, Ties that Binds**, © 2012, https://www.youtube.com/watch?v=tbNOiKKiNvA

91) Robert Scherrer, Disciplined Daydreaming: The Role of Ideas in Science and Science Fiction, **Analog** March / April 2019, pages (4 bottom)   4 – 9

SECTION 2
PART 1: THE FATE OF THE UNIVERSE

92) Cable Scharf, **The Zoomable Universe**, published by Scientific American – Farrar and Giroux, © 2017, pages 16 – 17

93) Changbom Park and Young-Rae Kim, Large–Scale Structure of the Universe as a Cosmic Standard Ruler, **The Astrophysical Journal Letters**, Volume 715, Number 2 May 17, 2010, The American Astronomical Society, pages 185 – 188, http://iopscience.iop.org/article/10.1088/2041-8205/715/2/L185/meta

94.1) Edward R. Harrison, **Cosmology**, published by Cambridge University Press, © 1981, pages 293–310

94.2) Charles Seife, **Alpha and Omega**, published by Viking, © 2003, pages 105, 109

95) Edward R. Harrison, **Cosmology**, published by Cambridge University Press, © 1981, pages 301 – 303

96) Edward R. Harrison, **Cosmology**, published by Cambridge University Press, © 1981, pages 301 – 303

97.1) Alex Filippenko, Dark Energy and the Runaway Universe, November 14, 2013, **Talks at Google**, https://www.youtube.com/watch?v=Guvv5olLxCQ

97.2) Carl Sagan, **Cosmos**, published by Random House, © 1980, pages 265 – 267

97.3) Matt Parker, **Things To Make and Do In the Fourth Dimension**, Farrar, Straus and Giroux, © 2014, page 341, 442

98) Dave Rothstein, © 1997 – 2018, What Is the Universe Expanding Into (Intermediate), **Ask an Astronomer**, pages 1 – 4, http://curious.astro.cornell.edu/about-us/104-the-universe/cosmology-and-the-big-bang/expansion-of-the-universe/623-what-is-the-universe-expanding-into-intermediate

99)   Yu L. Bolotin and L. V. Tanatarov, ***A Thousand Problems in Cosmology: Horizons***, ©
       2014, arXiv:1305.0259, arXiv:1310.6329v3 [physics.ed-ph] 9 Jan 2014, pages 1 – 61, https://
       arxiv.org/pdf/1305.0259.pdf

100)   Matthew R. Francis, How Big Is The Universe? (Universe In A Box, Part 4), © November 30,
       2011, ***Galileo's Pendulum.mht***, https://galileospendulum.org/2011/11/30/how-big-is-th
       e-universe-universe-in-a-box-part-4/

101)   Matthew R. Francis, How Big Is The Universe? (Universe In A Box, Part 4), © 2011, ***Galileo's
       Pendulum.mht***, https://galileospendulum.org/2011/11/30/how-big-is-the-universe-univers
       e-in-a-box-part-4/

102.1) Stephen Hawking, ***Black Holes and Baby Universes and Other Essays***, published
       by Bantam Books, published by Bantam Books, © 1993, http://www.google.com/
       url?sa=t&rct=j&q=&esrc=s&source=web&cd=19&ved=0ahUKEwic2cLrwuLb
       AhUI2oMKHZYKAaM4ChAWCFAwCA&url=http%3A%2F%2Fketabnak.com%2Fredirect.
       php%3Fdlid%3D68695&usg=AOvVaw1nh_dfRl6oNcDyLQyySMoz

102.2) Carl Sagan, ***Cosmos***, Episode 10: The Edge of Forever, © 1980

102.3) Alan H Guth, ***The Inflationary Universe***, published by Heilix Books and Addison–Wesley
       Publishing Company, Inc., © 1997, pages 245 – 252

102.4) Lee Smolin, ***The Life of the Cosmos***, published by Oxford University Press, © 1997, pages
       87–88

102.5) Brian Greene, ***The Fabric of the Cosmos Nova Transcripts "Universe or Multiverse?"***
       page 4, (http://www.pbs.org/wgbh/nova/physics/fabric-of-cosmos.html), © 2011

102.6) Clifford Pickover, ***Black Holes: A Traveler's Guide***, published by John Wiley and Sons
       Inc., © 1996, pages 136 – 138

103)   Edward R. Harrison, ***Cosmology***, published by Cambridge University Press, © 1981, pages
       271 – 279

104)   Christopher Pilot, ***Modeling Cosmic Expansion and Possible Inflation As A
       Thermodynamic Heat Engine***, © 2017, pages 1 – 19, https://arxiv.org/ftp/arxiv/
       papers/1705/1705.04743.pdf

105)   Edward R. Harrison, ***Cosmology***, published by Cambridge University Press, © 1981, pages
       311 – 328

106.1) Joao Magueijo, ***Faster Than the Speed of Light***, published by Penguin Books, © 2003,
       page 144

106.2) Lene Vestergaard Hau, *Scientific American Special Edition: The Edge of Physics*, May 31 2003, Frozen Light, pages 44 – 51

106.3) John D. Barrow, *The Book of Universes*, published by W. W. Norton & Company, © 2011, pages 274

107)   Joao Magueijo, *Faster Than the Speed of Light*, published by Penguin Books, © 2003, page 144

108.1) Lene Vestergaard Hau, *Scientific American Special Edition: The Edge of Physics*, May 31 2003, Frozen Light, pages 44 – 51

108.2) David Dugan, *Nova: Absolute Zero*, © 2008

109)   Istvan Szapudi, *Scientific American*, August 2016, Emptiest Place in Space, pages 28 – 35

110)   Walter Isaacson, *Einstein: His Life and Universe*, published by Simon and Schuster, © 2007, page 353

111)   Walter Isaacson, *Einstein: His Life and Universe*, published by Simon and Schuster, © 2007, page 353

PART 2: PROOFS OF A FOUR-DIMENSIONAL UNIVERSE DOWN TO ITS ATOMIC STRUCTURE8

Proof 1: The Expansion Rate of the Universe

112)   Anders Persson, "How Do We Understand the Coriolis Force?" *Bulletin of the American Meteorological Society*, Vol. 79, No. 7, July 1998, pages 1–13, https://journals.ametsoc.org/doi/abs/10.1175/1520-0477%281998%29079%3C1373%3AHDWUTC%3E2.0.CO%3B2

113.1) Dr. John P. Cise, Professor of Physics, KINEMATICS Units 4 & 5, *Cosmos Controversy: The Universe Is Expanding, but How Fast?* published by Austin Community College, jpcise@austincc.edu, February 20, 2017, http://cisephysics.homestead.com/files/ExpandingUniverse.pdf

113.2) T. Wayne, MRWAYNESCLASS.COM, *G's Felt In A Loop Example*, https://www.youtube.com/watch?v=qE4Teq5na6U

113.3) Charles A. Whitney, *Discovery of Our Galaxy*, published by Alfred a Knopf, © 1971, page 238

113.4) Govert Schilling, Constant Controversy, *Sky and Telescope,* June 2019, pages 22 – 29

114)   John D. Barrow, **The Book of Universes**, published by W. W. Norton & Company, © 2011, pages 86, 311

115)   Garry and Sylvia Anderson, **Space 1999: Episode 1: Break Away**, 1975

116.1) Edward R. Harrison, **Cosmology**, published by Cambridge University Press, © 1981, pages 283 – 292

116.2) Phillis Engelbert and Diana L Dupuis, **The Handy Space Answer Book**, published by Visible Ink Press, © 1998, page 375

117)   Edwin A. Abbbott, **Flatland**, published by Dover Publication Inc., © 1992 (1884), pages vii, viii

118.1) Richard Panek, **The Trouble With Gravity,** published by Houghton and Mifflin Harcourt, © 2019

118.2) Richard Panek, **Scientific American**, March 2020, A Cosmic Crisis, pages 30 –37

118.3) Licia Verde, Tommaso Treu, Adam G. Riess, **Tensions Between the Early and the Late Universe**, published by Kavli Institute for Theoretical Physics, © 2019, pages 1 – 10, https://arxiv.org/pdf/1907.10625.pdf?fbclid=IwAR1SQSJXNIe8d98RwJsd-7Kn3IILgqwolE1KRISQuJu-c21daCt8qHqCEO8

118.4) Dan Hooper, Is the BigBang in Crisis, **Astronomy: Special Issue**, October 2020, Pages 34 -- 43

119.1) Yu L. Bolotin and L. V. Tanatarov, **A Thousand Problems in Cosmology: Horizons**, © 2014, arXiv:1305.0259, arXiv:1310.6329v3 [physics.ed-ph] 9 Jan 2014, pages 1 – 61, https://arxiv.org/pdf/1305.0259.pdf

119.2) Neil deGrasse Tyson, **The Horizon Problem**, © 2014, https://www.youtube.com/watch?v=utuvHwg6Tvc

120)   Yu L. Bolotin and L. V. Tanatarov, **A Thousand Problems in Cosmology: Horizons**, © 2014, arXiv:1305.0259, arXiv:1310.6329v3 [physics.ed-ph] 9 Jan 2014, pages 1 – 61, https://arxiv.org/pdf/1305.0259.pdf

121)   Editor Peter Fraces, **DK The Universe: The Definitive Visual Guide**, published by DK Publishing Inc, 2005, page 489

Proof 2: Hyper Holes

122)   Clifford A. Pickover, **Surfing Through Hyperspace**, published by Oxford University Press, © 1999, page 50

123) Muriel Mandell, **Simple Science Experiments With Everyday Materials**, published by Sterling Publishing Company Inc., © 1989, page 72

124) Albert Einstein, On a Stationary System with Spherical Symmetry Consisting of Many Gravitating Masses, **Annals of Mathematics**, Second Series, Vol. 40, No. 4 (Oct., 1939), © 1939, pages 922 – 936, http://www.jstor.org/stable/1968902

125) Josh Lyle, How the 'Glory Hole' on Lake Berryessa Works / abc 10.com, **ABC 10 News, March 15, 2019**, https://www.youtube.com/watch?v=1pBiWghBSfl

126) Niayesh Afshordi, Robert B. Mann, and Razieh Pourhasan, **Scientific American**, August 2014, "The Black Hole at the Beginning of Time", pages 36 – 43

127) John D. Barrow, **The Book of Universes**, published by W. W. Norton & Company, © 2011, pages 76 – 80

128) Niayesh Afshordi, Robert B. Mann, and Razieh Pourhasan, **Scientific American**, August 2014, "the Black Hole at the Beginning of Time", pages 36 – 43

129) Henry Norman, **Just What Are Those "Singularities" Anyway?** Academia, © 2014, https://www.academia.edu/17451349/Just_What_Are_Those_Singularities_Anyway

130) John D. Barrow, **The Book of Universes**, published by W. W. Norton & Company, © 2011, page 129

131) Stephen Hawking, PBS Video **Stephen Hawking's Universe: Black Holes and Beyond**, © 1997

132) Theodore A. Jacobson and Renaud Parentani, **Scientific American** December 2005, An Echo of Black Holes, pages 69 – 75

133.1) Clifford Stoll, **A Hole in a Hole in a Hole – Numberphile**, Numberphile, September 8, 2016, https://www.youtube.com/watch?v=k8Rxep2Mkp8

133.2) Michael Spivak, **Comprehensive Introduction to Differential Geometry, Volume 1, Third Edition**, Publish or Perish, Inc, © 1999, page 22

133.3) B. H. Arnold, **Intuitive Concepts in Elementary Topology**, Prentice Hall, © 1963, pages 67, 68

133.4) Ian Stewart, **Flatterland**, Perseus Publishing, © 2001, pages 94 – 95

134) Chris McMullen, **Full Color Illustrations of the Fourth Dimension, Volume 2**, Amazon Kindle Direct Publishing, 2009, pages 33 – 35

135) John W. Macvey, **Time Travel**, Scarborough House Publishing, © 1990, page 123

136) John W. Macvey, **Time Travel**, Scarborough House Publishing, © 1990, page 123

137) David Freedman, **Discover** July 1990, Maker of Worlds, pages 46 – 52

138) Joanna Baker, **50 Physics Ideas: You Really Need to Know**, published by Quercus, © 2007, pages 32 -- 35

139) Lawrence S. Brown and Thomas A. Holme, **Chemistry for Engineering Students 2nd Edition**, published by Brooks/Cole, © 2011, pages 125 – 157

140) Grey Hautaluoma and Jennifer Morcone, and Megan Watzke, **NASA Death Star Galaxy Black Hole Fires at Neighboring Galaxy: Chandra – Exploring the Invisible Universe**, YouTube, December 17, 2007, https://www.youtube.com/watch?v=9POe3zFhE5g or https://www.nasa.gov/mission_pages/chandra/news/07-139.html

141) Editors at Time Life Books, **Voyages Through the Universe: Galaxies**, published by Time Life Books, © 1989, pages 70 – 71

142) Miguel Alcubierre, The Warp Drive: Hyper–Fast Travel Within General Relativity, **Classic and Quantum Gravity**, © 1994, pages 73 – 77, https://arxiv.org/pdf/gr-qc/0009013.pdf

143) T. W. Hart Quist, J.E. Dyson, D. P. Ruffle, **Blowing Bubbles in the Cosmos**, published by Oxford University Press © 2004, pages 36 – 45

144) Alan H Guth, **The Inflationary Universe**, published by Heilix Books and Addison–Wesley Publishing Company, Inc., © 1997, page 2

145) Alan H Guth, **The Inflationary Universe**, published by Heilix Books and Addison–Wesley Publishing Company, Inc., © 1997, page 2

146) Lee Smolin, **The Life of the Cosmos**, published by Oxford University Press, © 1997, page 87

147) Lee Smolin, **The Life of the Cosmos**, published by Oxford University Press, © 1997, page 87

148) Lee Smolin, **The Life of the Cosmos**, published by Oxford University Press, © 1997, pages 87–88

149) Lee Smolin, **The Life of the Cosmos**, published by Oxford University Press, © 1997, page 88

150) Lee Smolin, **The Life of the Cosmos**, published by Oxford University Press, © 1997, page 88

151) Lee Smolin, **The Life of the Cosmos**, published by Oxford University Press, © 1997, page 88

152) Lee Smolin, **The Life of the Cosmos**, published by Oxford University Press, © 1997, page 88

153) John D. Barrow, ***The Book of Universes***, published by W. W. Norton & Company, © 2011, pages 180 – 188

154.1) James Edward Biechier, ***Ten Minutes in the Dark***, © 2007, https://www.youtube.com/watch?v=bBP6dCkraJc

154.2) James Edward Biechier, ***The Physical Origins of Dark Matter and Dark Energy: Exploring Shadows on the Cave Wall***, Academia, © 2010, https://www.academia.edu/7673956/Physical_Origins_of_Dark_Matter_and_Dark_Energy_Exploring_Shadows_on_the_Cave_Wall?auto=download

155) Stephen Hawking, ***The Universe in a Nutshell***, published by Bantam Books, © 2001, page 118

156) Amir D Aczel, ***Pendulum: Leon Foucault and the Triumph of Science***, published by Atria Simon & Schuster Inc, © 2003, pages 91 – 98

Proof: 3 Chemical Reactions in Four Spatial Dimensions

157) Zulfikar Ahmed, ***Evidence and Argument for a Four Dimensional Spherical Universe***, August 28, 2012, pages 1 – 26, https://zulfahmed.files.wordpress.com/2012/08/universeiss45.pdf

158) Zulfikar Ahmed, ***Evidence and Argument for a Four Dimensional Spherical Universe***, August 28, 2012, https://zulfahmed.files.wordpress.com/2012/08/universeiss45.pdf

159) E.J. W. Whittaker, ***An Atlas of Hyperstereograms of the Four–Dimensional Crystal Classes***, Clarendon Press Oxford University Press, © 1985, pages 172–196

160) Chis McMullen, **Full Color Illustrations of the Fourth Dimension, Volume 2**, Amazon Kindle Direct Publishing, 2009, pages 30 – 32

161) P. Erik Gundersen, ***The Handy Physics Answer Book***, published by Visible Ink Press, © 1999, pages 159 – 171

162) Diego L. Rapoport, ***Möbius strip and Klein Bottle Genomic Topologies, Self-reference, Harmonics and Evolution***, Academia, December 18, 2016, https://www.academia.edu/15179783/M%C3%B6bius_strip_and_Klein_Bottle_Genomic_Topologies_Self-reference_Harmonics_and_Evolution, pages 1 – 61

163) James Edward Biechier, ***2008 Anomalous Physics***, © 2008, https://www.youtube.com/watch?v=cnjpDuPkEWA

164) Bertrand Toudic, Pilar Garcia, Christophe Odin, Philippe Rabiller, Claude Ecolivet, Eric Collet, Philippe Bourges, Garry J. McIntry, Mark D. Hollingsworth, and Tomasz Breczewski, ***Science*** January 4, 2008, Hidden Degrees of Freedom in Aperiodic Materials, pages 11, 41–42, 69–71

165) Wikipedia contributors, primary contributor Revision History Statistics, *Fluoroscopy*, published by Wikipedia The Free Encyclopedia, July 19, 2018, https://en.wikipedia.org/w/index.php?title=Fluoroscopy&oldid=849056833

166) Matt Parker, *Things to Make and Do in the Fourth Dimension*, published by Farrar, Straus and Gioux, © 2014, pages 209 – 210

167) Matt Parker, *Things to Make and Do in the Fourth Dimension*, published by Farrar, Straus and Gioux, © 2014, pages 209 – 210

168) The Science Elf, YouTube, *A Beginner's Guide to the Fourth Dimension*, June 20, 2016, https://www.youtube.com/watch?v=j-ixGKZlLVc

169.1) Joanna Baker, *50 Physics Ideas: You Really Need to Know*, published by Quercus, © 2007, pages 164–167

169.2) edited by Stephen Hawking (Albert Einstein), *A Stubbornly Persistent Illusion: The Essential Scientific Works of Albert Einstein*, published by Running Press, © 2007, page 173

169.3) Jeremy Bernstein, *Einstein,* published by Penguin Books, © 1978, pages 110 – 111

170) Larry D. Kirkpatrick and Gerald F. Wheeler, *Physics: A World View 4th Edition*, published by Harcourt College Publishers, © 2001, page 340–341

171) Larry D. Kirkpatrick and Gerald F. Wheeler, *Physics: A World View 4th Edition*, published by Harcourt College Publishers, © 2001, page 340–341

172) Wikipedia contributors, **NOEIN**, Wikipedia The Free Encyclopedia, © 2019, https://en.wikipedia.org/wiki/Noein:_To_Your_Other_Self, pages 1 – 5

173) Tom Jackson, *Philosophy: An Illustrated History of Thought*, Shelter Harbor Press, © 2014, page 129

174.1) Joanna Baker, *50 Physics Ideas: You Really Need to Know*, published by Quercus, © 2007, pages 120–123

174.2) edited by Stephen Hawking, *A Stubbornly Persistent Illusion: The Essential Scientific Works of Albert Einstein*, published by Running Press, © 2007, page 173

174.3) Jeremy Bernstein, *Einstein,* published by Penguin Books, © 1978, pages 110 – 111

175.1) Khan Academy, *Intuition About Simple Harmonic Oscillators - Physics*, July 29, 2016, https://www.youtube.com/watch?v=ZcZQsj6YAgU&t=52s

175.2) Every Think, YouTube, **How Do Atomic Clocks Work?** March 8, 2016, https://www.youtube.com/watch?v=l8CI3bs9rvY

176.1) Marshall Brian, **Marshall Brian's How Stuff Works**, published by Hungry Minds, © 2001, page 95

176.2) Every Think, YouTube, **How Do Atomic Clocks Work?** March 8, 2016, https://www.youtube.com/watch?v=l8CI3bs9rvY

177) Wikipedia contributors, **Resonators**, Wikipedia, The Free Encyclopedia, © 2019, https://en.wikipedia.org/wiki/Resonator, pages 1 – 6,

178) Joanna Baker, **50 Physics Ideas: You Really Need to Know**, published by Quercus, © 2007, pages 164 – 167

179.1) Veit Elser and Christopher L Henley, **Physical Review Letters**, December 23 1985, Volume 55, Number 26, Crystals and Quasicrystal Structures in Al-Mn-Si Alloys, pages 2883 – 2886, http://www.lassp.cornell.edu/clh/PUBS/alpha-85.pdf

179.2) T. Janssen, J. L. Birman, F. Denoyer, V. A. Koptsik, J. L. Verger–Gaugry, D. Weigel, A Yamamotto, S. C. Abrahams, and V. Kopsky, Foundations of Crystallography, Report of a Subcommittee on the Nomenclature of n – Dimensional Crystallography. II. Symbols for arithmetic crystal classes, Bravis classes and space groups, **Acta Crystallographica**, November 2002, Volume 58, Part 6, pages 605 – 621, http://scripts.iucr.org/cgi-bin/paper?S010876730201379X

179.3) R Veysseyre, D Weigel, T Phan, and J M Effantin, **Acta Cryst** (1984), A40, Crystallography, Geometry and Physics in Higher Dimensions. II. Point Symmetry of Holohedries of the Two Hypercubic Systems in Four – Dimensional Space, pages 331 – 337, https://onlinelibrary.wiley.com/doi/abs/10.1107/S0108767384000714

180) Hideo Nitta and Keita Takatsu, **The Manga Guide to Physics**, published by No Starch Press, © 2009, pages, 53 – 57

181) Hideo Nitta and Keita Takatsu, **The Manga Guide to Physics**, published by No Starch Press, © 2009, pages, 37 – 43

182) J. Divahar, 4 Dimensional Visualization, Mathworks, February 25, 2010, https://www.mathworks.com/matlabcentral/fileexchange/13503-4-dimensional-visualization

183) Matt Parker, **Things to Make and Do in the Fourth Dimension**, published by Farrar, Straus and Gioux, © 2014, pages 326 -- 327

184) Brian Greene, **The Elegant Universe**, published by Vintage Books a Division of Random House, © 2003, page 207 (hardback)

185) Eugene Khutoryansky, How to Draw 4,5,6, and 7 dimensions, **Physics Videos By Eugene Khutoryansky**, September 11, 2012, https://www.youtube.com/watch?v=Q_B5GpsbSQw

186.1) Wikipedia contributors primary contributor Revision History Statistics, **Time Crystal**, published by Wikipedia Free Encyclopedia, https://en.wikipedia.org/w/index.php?title=Time_crystal&oldid=850814936

186.2) Frank Wilczek, **Scientific American**, November 2019, Crystals in Time, pages 28 – 35

186.3) Trace Dominguez, **There's a New Form of Matter That Exists in Four – Dimensions**, D News, © 2017, https://www.youtube.com/watch?v=TjVi5L7cYLQ

Proof: 4 "Faster-Than-Light" Communications

187) Stephen Hawking and Leonard Mlodinow, **The Grand Design**, published by Bantam Books, © 2010, page 100

188) Stephen Hawking and Leonard Mlodinow, **The Grand Design**, published by Bantam Books, © 2010, pages 100 – 101

189) Stephen Hawking and Leonard Mlodinow, **The Grand Design**, published by Bantam Books, © 2010, pages 101 – 102

190) Stan Gibilisco, **Understanding Einstein's Theories of Relativity**, published by Dover Press, © 1983, page 86

191) Paul Halpern, **Einstein's Dice and Schrodinger's Cat**, published by Basic Books, © 2015 pages 234 – 235

192) Leonard Nimoy, **Star Trek IV: The Voyage Home**, © 1986

193) Wikipedia contributors primary contributor Revision History Statistics, **Vulcan (Star Trek)**, published by Wikipedia Free Encyclopedia, https://en.wikipedia.org/w/index.php?title=Vulcan_(Star_Trek)&oldid=849976053

194) Miguel Alcubierre, The Warp Drive: Hyper–Fast Travel Within General Relativity, **Classic and Quantum Gravity**, © 1994

Proof: 5 Gravitational Lenses

195) John Gribbin, **In Search of the Big Bang**, published by Bantam Books, © 1986, page 78

196) John Gribbin, **In Search of the Big Bang**, by John Gribbin, published by Bantam Books, © 1986, pages 78 – 79

197) John Gribbin, ***In Search of the Big Bang***, by John Gribbin, published by Bantam Books, © 1986, page 120

198.1) James Edward Biechier, Three Logical Proofs: The Five – Dimensional Reality of Space – Time, ***Journal of Scientific Exploration***, Vol 21, No 3, © 2007 page 523 – 542, https://www.researchgate.net/publication/287589188_Three_logical_proofs_The_five-dimensional_reality_of_space-time

198.2) Wikipedia contributors primary contributor Revision History Statistics, **William Kingdom Clifford**, published by Wikipedia The Free Encyclopedia, February 21, 2018, https://en.wikipedia.org/w/index.php?title=Wiliam_Kingdom_Clifford&oldid=826789941

199) James Edward Biechier, Three Logical Proofs: The Five – Dimensional Reality of Space – Time, ***Journal of Scientific Exploration***, Vol 21, No 3, © 2007 page 523 – 542, https://www.researchgate.net/publication/287589188_Three_logical_proofs_The_five-dimensional_reality_of_space-time

200) James Edward Biechier, Three Logical Proofs: The Five – Dimensional Reality of Space – Time, ***Journal of Scientific Exploration***, Vol 21, No 3, © 2007 page 523 – 542, https://www.researchgate.net/publication/287589188_Three_logical_proofs_The_five-dimensional_reality_of_space-time

201) Rudy Rucker, ***The Fourth Dimension: A Guided Tour of Higher Universes***, published by Houghton Mifflin Company, © 1989, pages 82–83

202) Joanna Baker, ***50 Physics Ideas: You Really Need to Know***, published by Quercus, © 2007, pages 164 – 167

203) Albert Einstein, ***Ideas and Opinions***, published by Crown, © 1954 page 348

204) Larry D. Kirkpatrick and Gerald F. Wheeler, ***Physics: A World View 4th Edition***, published by Harcourt College Publishers, © 2001, page 321

205) Larry D. Kirkpatrick and Gerald F. Wheeler, ***Physics: A World View 4th Edition***, published by Harcourt College Publishers, © 2001, page 321

206) Larry D. Kirkpatrick and Gerald F. Wheeler, ***Physics: A World View 4th Edition***, published by Harcourt College Publishers, © 2001, page 321

PART 3: EMBEDDED TIME TRAVEL

207) Stephen Hawking, PBS video ***Stephen Hawking's Universe: Black Holes and Beyond***, © 1997

208) Lawrence Krauss, ***The Physics of Star Trek***, published by Basic Books a Division of Harper–Collins Publishers, © 1995, page 13

PART 4: MOTION ORIENTED TIME TRAVEL

209) Donald Kingsbury, **Psychohistorian Crisis**, published by Tor a Tom Doherty Associates Book, © 2001, pages 87 –88

210) Donald Kingsbury, **Psychohistorian Crisis**, published by Tor a Tom Doherty Associates Book, © 2001, pages 286 – 287

211) Daniel Schroeder, **Physics Stack Exchange**, September 10, 2017, https://physics.stackexchange.com/questions/356412/what-does-enthalpy-mean/356432

212) John D. Barrow, **The Book of Universes**, published by W. W. Norton & Company, © 2011, pages 208 – 212

213) Khan Academy, June 13, 2011, **Enthalpy: Thermodynamics: Chemistry**, https://www.khanacademy.org/science/chemistry/thermodynamics-chemistry/enthalpy-chemistry-sal/v/enthalpy

214) Joanna Baker, **50 Physics Ideas: You Really Need to Know**, published by Quercus, © 2007, pages 120 – 123

215) Edward R. Harrison, **Cosmology**, published by Cambridge University Press, © 1981, pages 275 – 276

216) Alex Filippenko, Dark Energy and the Runaway Universe, November 14, 2013, **Talks at Google**, https://www.youtube.com/watch?v=Guvv5olLxCQ

217) Ron Cowen, **Science News**, August 21, 2004, Cosmic Melody, pages 119 – 120

218) Wikipedia contributors, **Mechanism of Sonoluminescence**, published by Wikipedia The Free Encyclopedia, December 6, 2019, https://en.wikipedia.org/w/index.php?title=Mechanism_of_sonoluminescence&oldid=929573253

219) Lawrence Krauss, **Quintessence**, published by Basic Books, © 2000, page 31

220) Ron Cowen, **Science News**, August 21, 2004, Cosmic Melody, pages 119 – 120

221) Gerald E. Tauber, **Man and the Cosmos**, published by Greenwich House, © 1979, pages 122 – 129

SECTION 3: TECHNOLOGY APPLIED SCIENCE
PART 1: WARP DIVE

222) Miguel Alcubierre, The Warp Drive: Hyper–Fast Travel Within General Reltivity, **Classic and Quantum Gravity**, © 1994, pages 73 – 77, https://arxiv.org/pdf/gr-qc/0009013.pdf

223) Miguel Alcubierre, The Warp Drive: Hyper–Fast Travel Within General Relativity, *Classic and Quantum Gravity*, © 1994, pages 73 – 77, https://arxiv.org/pdf/gr-qc/0009013.pdf

224) Miguel Alcubierre, The Warp Drive: Hyper–Fast Travel Within General Reltivity, *Classic and Quantum Gravity*, © 1994, pages 73 – 77, https://arxiv.org/pdf/gr-qc/0009013.pdf

225) *NASA's Warp Speed Starship Prototype IXS Enterprise Great CGI Animation*, November 9, 2016, https://www.youtube.com/watch?v=hc8vAJHpw8o

226) Mae Jemison and Dana Meachen Rau, *The 100 Year Starship*, published by Scholastic Inc., © 2013

227) Miguel Alcubierre, The Warp Drive: Hyper–Fast Travel Within General Relativity, *Classic and Quantum Gravity*, © 1994,pages 73 – 77, _

228) Ethan Siegel, *Star Trek Treknology*, published by Voyageur Press, © 2017

SECTION 3, PART 2: FOUR-DIMENSIONAL FASTER-THAN-LIGHT COMMUNICATIONS

229) Stan Gibilisco, *Understanding Einstein's Theories of Relativity*, published by Dover Press, © 1983, page 86

230) Graham A. J. Worthy and Cherie L. Yestrebsky, *Scientific American*, October 2018, Break Down Silos, pages 64 – 67

231) Jim Daley, *Scientific American*, January 2020, Quantum Loop, page 18

232) Urasi Sinha, *Scientific American*, January 2020, The Triple Split Experiment, pages 57 – 61

233) Hideo Nitta, Masafumi Yamamoto, and Keita Takatsu, *The Manga Guide to Relativity*, published by No Starch Press, © 2011, page 113

234) Hideo Nitta and Keita Takatsu, *The Manga Guide to Physics*, published by No Starch Press, © 2009, page 40

ABOUT THE AUTHOR

Roger I. Parker II has been fascinated with physics since boyhood. He is also an avid reader and enjoys sharing books that he loves with his young niece. A descendant of Betsy Ross, he has worked on dozens of research projects throughout his life, including this book. He was born and raised in Wilmington, Delaware, where he currently lives.

Printed in the United States
By Bookmasters